Charles Johnson

Catalogue of interlocking and railroad signaling appliances

Charles Johnson

Catalogue of interlocking and railroad signaling appliances

ISBN/EAN: 9783741112676

Manufactured in Europe, USA, Canada, Australia, Japa

Cover: Foto ©Andreas Hilbeck / pixelio.de

Manufactured and distributed by brebook publishing software
(www.brebook.com)

Charles Johnson

Catalogue of interlocking and railroad signaling appliances

THE JOHNSON RAILROAD SIGNAL COMPANY

RAHWAY, N. J.

NEW YORK OFFICE, 146 BROADWAY

CHICAGO OFFICE, THE ROCK ISLAND DEPOT BUILDING

CHARLES R. JOHNSON - PRESIDENT AND GENERAL MANAGER

WALTON O. KERNOCHAN - - - - - TREASURER

HENRY JOHNSON - - - - - MANAGER OF WORKS

GEORGE E. READ - - - - - - - SECRETARY

CATALOGUE

OF

INTERLOCKING AND RAILROAD SIGNALING APPLIANCES

PREFACE

THE credit of the introduction of Interlocking Signals in the United States is due to Messrs. Toucey & Buchanan, of the New York Central & Hudson River Railroad, the former, General Superintendent, and the latter Superintendent of Motive Power. These two gentlemen very early saw the advantages of concentrating switches as much as possible so that they could be worked from a central point and protected by a much smaller number of Signals than would be necessary if a Signal was placed for every switch. They devised an Interlocking machine, and the first one was fixed at Spuyten Duyvil Junction, in New York city, in 1874, and remained in service until 1888. This machine compares very favorably with the earlier machines used in England which was the birthplace of Interlocking Signals.

The Pennsylvania Railroad was also early in the field but they sent to the well known firm of Saxby & Farmer of London, England, for a complete machine, etc., with Signals and connections, which were sent over and fixed at East Newark Junction on the New York Division, being put into service on February 11, 1875 where they are now working.

In 1876 Messrs. Saxby & Farmer sent a very complete model of their system of interlocking and block signals, to the Centennial Exhibition in Philadelphia. This exhibit did more perhaps than any other one thing to acquaint railroad managers in this country with the systems of interlocking and block signals, which were then extensively used on European Railroads.

Very shortly after this, the Elevated Railroads of New York were built and equipped at the most important points with the Saxby & Farmer machine manufactured by the Jackson Manufacturing Company of Harrisburg, Pa., who had purchased the patent rights for this country.

The slow progress of the introduction of Interlocking up to the year 1887, was due partly to a want of knowledge of its advantages and partly to the serious depression of trade which was felt so long. The *Railroad Gazette* and other Railroad papers, did their utmost to urge Railroad Officers to take up and investigate the subject and did much to point out its uses and advantages, but even at the present time there are still comparatively few railroad engineers who have a thorough knowledge of the latest improvements in interlocking and block signaling.

In this preface the writer can do no more than point out the chief merits of the Interlocking system.

The advantages of Interlocking may be classed under two heads:

I.　Increased safety.

II.　Increased facility in the handling of traffic at busy points.

Increased safety is assured by working each system of switches and signals from a central point, the mechanism for operating such system being so arranged that so long as the apparatus is kept in good condition, movements taking place under the sanction of the operator, as expressed by the lowering of a signal or signals—and no other movements can take place without such sanction —are secure against collision from conflicting directions, and disturbance of the switches traversed by such authorized movements.

Much of the increased facility that may be obtained depends upon a good lay out of tracks as well as of the Signals. When switches are arranged so as to obtain the greatest number of movements with the least possible running, much more rapid handling of trains can be obtained with safety by one man who handles all switches and Signals from an elevated point, where he can see each movement and anticipate each requirement, than by a number of men who have to run from switch to switch to throw them.

One can best realize this by watching the movements of trains at Grand Central Station, New York; Broad Street Station, Philadelphia, or the Boston Yard of the Boston & Albany R. R., and comparing them with those at other busy places which have no Interlocking.

The introduction of Interlocking Signals and switches has not always been a complete success, for obvious reasons, viz.: faulty arrangement of tracks and incomplete signaling. The existence of either or both these conditions will, of course, mar the complete efficiency of any system, and if from desire to economize or other reasons, systems having such faults are introduced or allowed to remain, satisfactory results can scarcely be expected.

It is very satisfactory to all concerned in good Signaling to know that an increasing number of roads are alive to the importance of having a properly equipped Signal Department. Several important roads have appointed Signal Engineers whose duty it is to superintend the construction of new work and maintain it in good condition afterwards.

In making these appointments it is well to secure the services of men of some years' experience, because, although Signaling may appear to be very simple and is so in small plants, there are so many intricacies in large ones (and every road will have some such), that nothing but long and varied experience can fit a man to fill the position satisfactorily.

There is something about signals very fascinating to the inventive faculty, and the application for patents, and the patents granted for Signaling devices must approximate very nearly to the infinite number of car couplers and rail sections. While it is very laudable to desire to improve Signaling devices, it must be remembered that there is great economy in uniformity, and changes should not be made unless some decided advantage is gained.

The experience of twenty-five years has pretty conclusively shown among other things that the Semaphore Signal is the most satisfactory type of signal; that switches and locks should be worked by pipe; that facing switches should be fitted with facing point locks; that facing point locks should be duplex, *i. e.*, so arranged that in the event of the breakage of connections, the plunger of the lock cannot be thrown into the wrong position of the switch; that two lines of wire should be used to each signal; that signal blades should be so constructed as to go to the danger position in case of breakage of connections anywhere between the operating lever and blade; that wires to distant signals should be automatically compensated; that iron plates should be fixed under switch points to keep the track accurately to gauge; that plungers of facing point locks should not be pointed; that cranks and pipe compensators should be fixed on foundations firmly embedded in concrete; that all side tracks connected to main tracks should be "trapped," *i. e.*, have a derailing switch to prevent cars coming on to the main track until the switch is set for the side track; that a signal shoud be given for every train movement; that high signals should only be used for main running tracks; that separate signal posts should be used for each track running parallel or converging; that one post with one or more blades (various systems are in use for indicating the route open) should be used for diverging tracks; that it is a most dangerous and reprehensible practice to displace or disconnect any part of safety appliances such as detector bar, switches, switch locks, machine interlocking, except in cases of absolute necessity, and then only temporarily and under proper protective conditions, such as pad Jocking the switches affected, issuance of caution notice and employment of flagmen at the positions of danger; that all ground connections should be well drained and all the appliances kept clean.

The points on a railroad where the most train movements are found are usually chosen to introduce Interlocking Signals, but an exception to this rule is found at grade-crossings and drawbridges, of which there are so many in this country. Most of the States some years since passed laws compelling Railroad Companies to bring their trains to a full stop before crossing a drawbridge or a grade-crossing. These laws have been found very irksome, not only on account of the cost of an unnecessary stop, but, from the delay caused by stopping passenger and heavy freight trains. This has been very clearly pointed out in some of the reports of Railroad Commissioners.

For a simple grade-crossing protected by 4 derailing switches, 4 home and 4 distant signals, the most simple form of Interlocking answers every purpose, and great efforts have been made to reduce the cost to as low a figure as possible. It must be remembered, however, that no matter how simple the interlocking, it should be arranged to be perfectly safe under all circumstances, and easy to maintain in good condition so that one man will be able to properly maintain several crossing towers. In the struggle to introduce cheap appliances, too little attention has sometimes been given to a proper factor of safety. There are crossings now being used with the derailing switches worked without a facing point lock or detector bar. This should absolutely be prohibited, as innumerable wrecks have occurred through the throwing of a switch under a train, and one of the most disastrous accidents that ever occurred was caused in this manner before, however, the facing point lock was in general use. It must not be forgotten that a derailing switch is a facing switch, which ought to be avoided as much as possible, consistent with proper handling of traffic. One of the first conditions, then, in connection with derailing switches should be to make it impossible to give a clear Signal with the switch open or partially open. With ordinary Interlocking this is improbable but not impossible. The only absolutely certain method is by working the Signal by means of the last movement of the plunger of facing point lock as shown on page 68. A switch detector, however, worked by the home signal connection may well be accepted as sufficiently certain, but without this a crossing should not be considered absolutely safe. In this connection we may consider the working of switches and locks by two lines of wire which is far less costly than pipe, but which is open to objections that should be clearly stated and understood. It has been demonstrated that switches and locks can be worked by means of continuous wire and they are certainly easier for the operator. But it is equally certain that perfect means have not yet been found to automatically counteract the effects of the stretching of the wires caused by varying strains. Adjusting screws are furnished and answer perfectly well when handled by competent men, but just at those places where wire working switches are likely to be used, competent repair and maintenance men are least likely to be found. While, therefore, we are prepared to connect switches by wire, and have the most perfect appliances in the field for so doing, we recommend the pipe connections solely because they are more easily kept in order. The wire working can be made perfectly safe, except that, when from any cause they become too slack to throw the switch, the operator is usually not sufficiently skilled to know how to tighten them to do their proper work. A passenger train may be standing waiting for the signal which the operator is unable to lower, owing to the imperfect action of the switch. He becomes excited, and instead of going to the switch to ascertain the trouble, he will wave his lantern for the engineman to come ahead, which the latter will frequently do, and so derail his train. This has happened several times.

A great many efforts have been made to work and lock a switch by means of one lever, and various devices are in existence for accomplishing that purpose. We believe we have the only movement that does the work perfectly and in a thoroughly satisfactory manner. We have accomplished this by giving a long initial stroke to the pipe connection which thereby reduces the power of the operator for rupturing them by giving him less leverage. At the same time, by using our anti-friction pipe carriers, the force required to move the connection is much lessened, and finally the switch and lock movement itself is so designed as to give the minimum of resistance in its proper work, and the maximum for rupture. With this device we claim that switches can be worked with greater facility than those having a separate lock lever, and as safely. When a connection becomes broken the operator knows it through his switch detector, which prevents the signal from being lowered unless the switch is properly home.

Our devices for working signals and switches from the centre of a drawbridge are now so complete, that not only do we obtain as perfect working as from ordinary towers, but no trouble is experienced from the changing position of the draw due to expansion and contraction, and the movement caused by passing trains. It very frequently happens that switches are located near the end of a drawbridge, and the use of a machine fixed in the centre of the draw and worked by the draw tender saves the expense of switchmen at one or both ends of the draw. It is very essential, however, that the connections be so arranged as to require little adjustment and be easily kept in good order. Due allowance should also be made for the jarring to bridge couplers, caused by trains passing on and off the bridge. All these requirements are met by the appliances shown in our illustrations.

A fruitful source of danger to trains is the misplaced switch which is continually causing disaster and which can almost invariably be avoided by the use of distant switch signals. It is absolutely certain that with facing switches unprotected by a signal these accidents will continue to happen in the future, as they have done in the past. It is not an expensive matter to have these signals and they can be arranged to be fixed in connection with any kind of switch stand. We have been unable in this issue of our catalogue to illustrate distant switch signals, but will be pleased to furnish plans and prices upon application.

For roads not having sufficient traffic to warrant the use of distant switch signals we can furnish padlocks for the ordinary switch levers so arranged that the switchman cannot take out the key of the padlock until the switch is set and locked for the main track. For switches also that are too far from a tower to be conveniently worked, we have a key locking arrangement, by means of which, a key must be taken from the tower to open the outlying switch, and until the key is brought back to the tower no signal can be given for a train to proceed in the direction of the switch, and of course the key cannot be brought back until the switch is set and locked for the main track. This method is a slow but very safe arrangement.

It very frequently happens that a signal tower is located at or near a street crossing, in which case it is decided economy to work the gates from the tower, and they may be interlocked with the signals or not, as may be found most desirable. This is very often found much more convenient, as well as safer, than having a separate man on the ground, who is liable to raise or lower his gates at the wrong moment, and besides, cannot see approaching trains so well as the man in an elevated tower. The ordinary lifting gates may be used, or swinging gates which close against the street in one position, and against the railroad in the other, so preventing cattle, etc., from getting on to the railroad when being driven over the crossing.

Various devices are in use for notifying enginemen of the position of signals during foggy weather. The most usual method of doing this is to place men at the signals with torpedoes, which they fasten to the rails according to the position of the signal. Unless this is done, or some automatic system used, trains will necessarily be delayed. So far as we know, nothing has yet been put into service that gives complete satisfaction, although numerous inventions have been made.

It is the custom in France to attach a torpedo to each home signal, so arranged that when the signal is at "danger" the torpedo is on the track, and when the signal is at "clear," the torpedo is withdrawn clear of the track, so that only when an engine or train runs past the signal at danger is the torpedo exploded. It is often important to know if an engineman has overrun his signal and this will give some indication, but not certain evidence, as there is nothing to prevent an operator throwing his signal to "danger" during the passage of a train, and so putting the torpedo on the track in front of the wheels.

The Palmer torpedo signal is in use to some extent and has given, so far as we know, general satisfaction. It works with the home signal as described above, but the instrument is arranged to hold five torpedos, and when one is exploded another takes its place until the five are exhausted, when the box has to be filled again.

Some efforts have recently been made to introduce an illuminated blade for signals, so as to show a night signal as near as possible like the day signal, but so far these efforts have not been very successful. The idea of illuminating the blade is quite an old one, and has been extensively tried, but never with enough success to displace the usual lamp showing red for "danger" and white for "all clear." Notwithstanding that some objections can be raised to this method of night signaling the fact remains that if accidents do occur through its use they rarely or never come to light and tens of thousands of these signals are in service and have been for years. It is quite probable that could equally good results have been obtained by colors for day signals, color instead of position would in all probability have been chosen. With a sky background the position signal by day shows perfectly, but unfortunately we cannot always obtain a sky background, so that it is impossible to give an equally good signal for all places. Ordinary observation will convince any one that a day signal may much more easily be passed unobserved than a night signal. The improvement needed then, is not in night but in day signals.

In considering the question of position signals for night, we need to be careful not to be allured into its adoption for the sake of comforming to the principle adopted for day signals. It is quite reasonable to have one principle for day and another for night; and unless it can be shown that there are important grounds for change other than mere uniformity, it would seem to be undesirable to make a new departure from an old established system which has worked so well in the past.

What are the advantages to be gained by introducing illuminated blades instead of the different colored lights, that would warrant railroad companies to depart from present usage? They perhaps are more distinctive, but certainly are not so arrestive as a strong light through a good red lense. Are they less expensive in first cost or to maintain? On the contrary, in both they will exceed the present methods. Will they retain the same uniform state of illumination? No; they are more liable to derangement, and more difficult to keep from becoming dim and obscure. Are they more desirable because of color blindness? No; color blindness can readily be detected and precaution should be taken to remove men from a position they are unfitted for. Trains carry colored lights in their rear and the misunderstanding of such would lead to accidents.

It may be said that their chief advantage lies in the fact that they differ from other lights about cities, more than the old systems.

Let us examine this. What is there in it? On the surface it is striking. Are not enginemen conversant with the road they run over? do they not know the location of all their signals, and can a light be added to or taken from the systems through which they pass without being detected by them? If so, we are in a sorry plight, because neither system gives us security. Lights may be extinguished and the rule which says, "*The absence of a signal where there should be one must be taken as a danger signal*" presupposes knowledge of the positions of all signals which govern the movements of the enginemen, and if one fails to observe the absence of a signal where there should be one, great risk is certainly incurred. Where is the danger in additional lights along the route either transitory or fixed? If of the same color as the railroad's safety light it can only be misleading through the extinction of the proper signal light and occupying its position. If of the same color as the railroad's danger signal the worst that could happen would be slight delay. The idea of street lights being mistaken for signal lights by a railroad engineman is far fetched and imaginary and does not rest upon good grounds.

As the proper working of any kind of mechanism depends to a great extent upon the condition in which it is kept we have appended a few general rules for maintenance which have received very general approval.

In our mechanical devices we shall endeavor to keep to one standard as much as possible consistent with a due regard for decided improvements.

GENERAL INSTRUCTIONS FOR OPERATORS AND FOR THE MAINTENANCE AND REPAIR OF INTERLOCKING MACHINES AND WORK IN CONNECTION THEREWITH

1. See that no obstacle exists on a route before setting the signal for it.
2. The normal position of signals is at "danger."
3. When no movement is immediate keep signal levers in normal position.

4. Never move a switch when a train covers it.

5. Never move a detector bar when a train covers it or is closely approaching it.

6. Put signal to danger as soon as train has passed it.

7. When signal has once been given to a train, should it be necessary to change the position of signals or switches, the signal may be changed to danger, but no signal must be given to a train on a conflicting route until the train which first had the signal has come to a full stop.

8. In case of accident notify proper officer.

9. Report any failure of lamps or irregularity of any kind to the proper officer.

10. Report to the proper officer any failure of interlocking.

11. Repairmen to obey the instructions for their guidance.

12. In case of a " run off," pass no more trains until you are satisfied that all parts liable to damage are in proper order.

13. In case a distant signal from any cause shows " clear " when the home is at " danger," fasten the distant signal in the shortest possible time to " caution " and discontinue its use until repaired.

14. Use a flagman to protect disconnected switches.

15. Allow no one to use tools on the machine except the regular workmen.

16. Do not allow the slightest change in locking except by a workman having a written permit from proper officer.

17. Permit no one to enter Tower whose duty does not require him there.

18. During cold weather move levers frequently to prevent connections from freezing up.

19. All parts of interlocking machinery are made fully as strong as the work of the machine requires.

20. Levers should be handled with a steady movement, and not in such a way as to indicate that the main object of the operator is to break the machine if possible.

21. Inspect switches, signals and locks daily.

22. Keep all switch points clear of cinders, ballast, etc.

23. Keep lamps clean. Light them and place in position at the proper time.

24. Keep the Lever Tower clean, together with all machinery in same.

25. Oil all parts of the interlocking requiring it, being careful to wipe off all old oil to prevent gumming up of working parts.

26. Keep fire buckets filled and ready for use in case of fire, and use them for no other purpose.

27. Never leave a detector bar disconnected without a flagman to protect the switch.

28. Report every case of an engineman running past a stop signal at " danger."

BLOCK SIGNALING

It has been demonstrated beyond doubt that safety in the running of trains cannot be obtained without the use of the block system, but the question of what is the best kind of block signaling and to what extent Railroad Companies can introduce what is certainly a costly system, either in first cost or in running expenses, is not by any means proved. If automatic signaling could be relied upon always it would probably be accepted as the most satisfactory and would certainly be the least expensive; but unfortunately it cannot always be relied upon, and when it does fail there is no one to tell an engineman that it has failed. There are automatic signals no doubt that show a very good record but they are by no means without failures. It is comparatively easy to construct and fix signals that in case of failure go to the danger position, but as soon as this happens with any frequency enginemen regard the signal with suspicion and either run so slowly as to lose time or run with reckless disregard of what the signal shows. As we do not make a specialty of automatic block signals we must leave this question with the above expression of our opinion with regard to them.

Block Signaling worked by operators is in much more general use, and there are various devices by means of which signals are transmitted from one operator to another. In the United States the ordinary telegraphic instruments are in most general use, and in this case operators ask, and are told, if the track is clear, and when so told lower their signal for a train to proceed. This system is sufficiently elastic to suit absolute or premissive block. Except where the traffic is very dense and towers can be placed at short intervals, absolute block, while giving complete safety, is felt to be irksome, and for that reason the permissive block is allowed to be freely used, notwithstanding that it is continually proving itself far from safe.

Absolute block as understood in Great Britain and Europe is scarcely known in this country. There absolute block means that before "line clear" can be given for a following train, the preceding one must be protected by a distant signal, a home signal and a starting signal, so that before a rear collision can occur a following train must run past one "caution" and two "stop" or " danger " signals. In other words an operator is only allowed to give back "line clear" when the last car of the train protected has passed beyond his starting signal. The great majority of block towers in this country have neither distant signals nor starting signals. There being one home signal for each direction usually fixed on one post at the tower itself. Generally speaking these block signals are located at passenger stations, and when that is so it is much better to have separate posts fixed at each end of the station so that in the event of the section ahead being blocked a train can draw into the station and do its regular business without passing a signal at "danger," when in most cases the block ahead will have been cleared.

There are various special instruments in use for block working designed to give operators the least possible chance for making an error, of which the best known are those invented by Tyer, Preece & Spagnoletti. None of these have been introduced into this country, but as the block system becomes better known they are quite likely to be used.

It has been found, however, that unless some means are employed to check both the receiver and sender of a train by each other and by the passage of the train itself, collisions, although rare, will sometimes occur from the mistakes of operators. This has been accomplished by the union of the Interlocking and Block systems whereby the signal for a train to proceed into a section cannot be lowered until the operator at the station ahead has given permission, and he on his part cannot give permission until the preceding train has passed his block signal and he has thrown it to "danger." Mr. Sykes' system has accomplished this in the most satisfactory manner and it is in service on the New York & Harlem Railroad, the N. Y. N. H. & H. R. R. and the N. Y., L. E. & W. R. R. to a considerable extent.

We regret that in this issue of our catalogue we are unable to fully illustrate and describe Mr. Sykes' system, but we shall do so in a later edition, and in the meantime will be pleased to furnish full particulars upon application.

NOTE

We have so arranged our Catalogue that in ordering any of the appliances, or parts of same, it will only be necessary to give the Numbers or Letters in the margin of the page.

THE JOHNSON INTERLOCKING MACHINE

Pat. No. 317137, March 19, 1885
"   "   392734, Aug. 17, 1888

SIDE ELEVATION   FRONT ELEVATION

THE JOHNSON INTERLOCKING MACHINE

THIS machine was designed to avoid certain defects in other interlocking machines, and to give a simple, strong and easily accessible locking.

The locking system is one of the oldest, the Stevens, but is actuated by the latch rod. All the locking is arranged in a single tier, and in a vertical plane, making examination of the locking very easy. There are only three styles of locking dog, and these accomplish, very simply, all ordinary and special locking. Any part of the locking may be removed or altered without disturbing locking having no relation to the alteration. The various wearing parts are of cold rolled iron and steel. As regards the latch actuation, we claim that this machine has the simplest and most durable movement extant. The Johnson Machine has a considerable advantage over other machines in the accessibility of the locking for repairs or changes, and in the simple and strong form of the locking dogs. It is generally acknowledged that the locking should be actuated by the preliminary action of the spring latch rod, and one of the most important reasons for this conclusion is that with direct attachment of the locking to the lever, it is often difficult to determine, when a lever cannot be moved, whether the working connection or the locking is holding it. In busy places, where the operator is in a hurry, unnecessary strain is often brought to bear on the locking in such a case. The plate shows a sectional side elevation, and a front elevation, of a four-lever Johnson machine ; 1, 2, 3, and 4 are wrought-iron levers centred on a girder 2, attached to legs 1, 1. The stroke of these levers is limited by portions of the segments 7, 8, 9, which form in combination with the spring latch 17, the well known means for holding the lever in either its home or its reversed position. The segments are carried by front and back girders 6 and 5, which in turn are supported at their ends by the beams A, H, and braced by being bolted to the beam A, J. The three girders are made for spans of 4 and 8 levers. The switch rods are connected to the levers at A, C. The gain stroke lever Z being used for wire connected signals only. It will be

THE JOHNSON INTERLOCKING MACHINE

Pat. No. 317137, March 19, 1885   Pat. No. 392734, Aug't 17, 1888

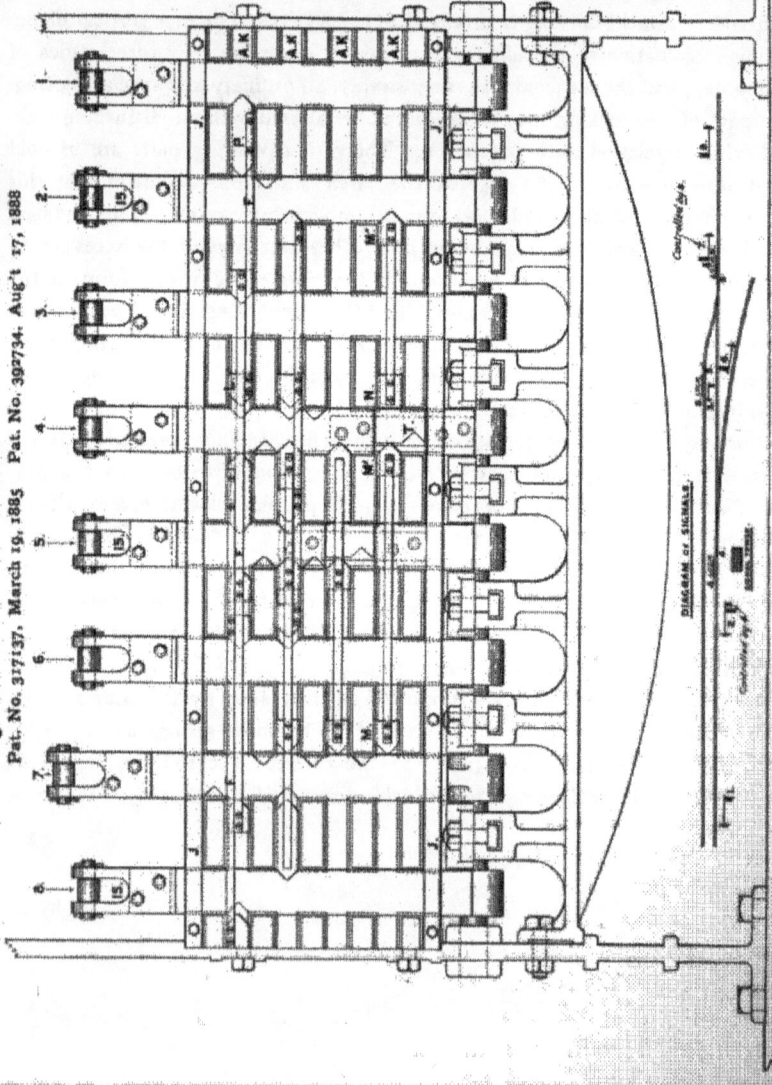

ELEVATION OF LOCKING

seen that the interlocking is all beneath the floor level, and is easy of access, as that portion of the floor, which is adjacent to the windows, and rests at one end on the ledge 6' of the girder 6, is cleated and removable.

The active and silent movements of the latch rod are communicated to the locking tappet H by means of the connecting rod G, and small reversing rocker 13, centred at 13' to the brackets 11 and 12, which are bolted by turned bolts in reamered holes on the main lever. The locking tappets are connected to the reversing rocker by a friction roller, which fits the curved slot in the rocker, and is centred by the jaw 15. If the tappet H were locked in the position shown, it will be readily seen that it would be impossible to raise the latch by pulling the latch handle 447.

In case the tappet is free, the intention of moving the main lever, as expressed by grasping the handle, and raising the latch, will raise the tappet and effect all the locking of other lever latches necessary to the safe movement of the lever in question. This movement also brings the curved slot in the rocker 13, radial to the centre of the main lever, so that the result of reversing the lever is a silent action upon the locking tappet. As the latch is dropped in the reversed position of the lever, the tappet H is raised further, and effects the necessary releasing of those levers which should be released when that lever is reversed.

The action of one tappet is made to release or lock other tappets, as the case may be, by transverse connections and dogs, carried by the locking plate 4, which also serves to guide and retain the tappets. By reference to the illustration this will be seen more clearly. Here is shown a front view of a locking plate for eight levers, of which Nos. 1 to 8 are the tappets, of cold rolled iron, free to slide vertically in planed recesses of the cast-iron locking plate, and retained by wrougth-iron strips, J, J. All the tappets are shown in their home positions except No. 7 which is reversed. The malleable jaws 15 carrying the friction rollers, are screwed to these tappets as shown. Transverse planed grooves A, K carry the cold rolled dogs L, P, etc.

These dogs are connected where necessary by the $\frac{1}{4}$ inch square cold rolled bars F, F, F, etc., which are fastened to the front of the dogs by small steel machine screws. By a recent improvement, three connecting bars may be used to each line of dogs, so that the locking requires less than half the space it formerly did when only one connecting bar could be used to each space.

The locking dogs with their connecting bars, are retained in their recesses by a plate and bolts shown at K.

The locking shown in our illustration applies to the safe working of a single line junction. The action of the dogs upon the tappets is so simple that the drawing will explain itself in this respect, and will be perfectly understood when compared with the locking sheet. It will be well, however, to draw attention to the method of performing the special or conditional locking. By reference to the locking sheet it will be seen that 2 locks 7 when 4 is home, but not when 4 is reversed. This lock is accomplished by M, M' T, N, M," in the following way. M, M' and N, M" are four dogs connected as shown. T is a transverse sliding section of the tappet 4, being rabbeted into the main tappet which has a gap at this point holding the slide T. The slide has a mitre notch, which, when 4 is home, comes opposite to the dog M', so that 2 may be raised together with 7 when 4 is thus, as the aforesaid notch of the slide T is simply made to coincide with the dog M', by the tappet 2 thrusting the dog M" outward. But suppose the tappet 4 is first raised, then the solid portion of the slide T will just fit between the dogs M' and N, thus forming a rigid connection between the dogs M' and M", making it impossible to have 7 and 2 raised simultaneously.

It will be noticed that this special locking is very simple ; all its parts being in the same plan, and on the same principle as the ordinary locking. By this method of special locking, any conditional lock may be performed. Although the locking is very rigid, when a latch is free, the movement of the locking offers slight resistance. The twist handle 16 will be supplied when preferred.

Estimates and Plans prepared for Interlocking any Junction, Grade-Crossing, Yard or Draw Bridge upon application.

The arrangement of the Levers and Table of Locking for the Junction shown in our Diagram of Signals on page 16 are given below.

LOCKING			LEVERS	
LEVER	RELEASES	LOCKS	NO.	FUNCTION
1.....	....	(2.) 4.	1.	Distant Signal.
2.....	1.......	4. (4.) 6. (7 when 4 is home.)	2.	Home Signal. (2 Blades.)
3.....	....	(5.) 7.	3.	Siding Signal.
4.....	5. 6.....	1. (2 B. S.)(7 when 5 is home.)	4.	Switch, Lock and Detector Bar.
5. ...	3.......	(4.) (7 B. S.) 8.	5.	2 Switches, 2 Locks and 2 Detector Bars
6.....	....	(4.) 2.	6.	Home Signal.
7.....	8.......	5. (5.) 3.	7.	Home Signal. (2 Blades.)
8... .	....	(7.) 5.	8.	Distant Signal.

THE JOHNSON

RAILROAD SIGNAL COMPANY'S

INTERLOCKING

PARTS OF MACHINE

INTERLOCKING MACHINE

The Johnson Machine has preliminary interlocking with all the well-known advantages. The peculiar advantages of this machine are as follows: The simplicity and durability of its parts. The extreme simplicity and ease with which all special locking is performed, any part of the locking mechanism may readily be removed for alteration or examination without disturbing locking, having no relation thereto. Although the locking is very rigid, when a latch is free the movement of the locking offers slight resistance.

Order No	
1	Side Stand (or Leg), only.
2	Bottom Girder only for carrying levers. These are made for sets of 4 or 8 levers. Our illustration shows an 8 lever girder.
97	Cap for bearing only (for lever centre pin).
4	Locking Plate only, for 8 levers and 12 locking bars. These are made for 4 or 8 levers and any required number of bars.
5	Top Front Girder only, to carry segments and bolt to stand.
6	Top Back Girder only, to carry segments and bolt to stand.
7	Left-hand End Segment only.
8	Centre Segment only.
9	Right-hand End Segment only.
172	End Plate, for right and left-hand segments only.
11	Left-hand Rocker Bracket only.
12	Right-hand Rocker Bracket only.
13	Rocker only.
15	Rocker Tappet Jaw only.
16	Loop or Swivel Handle only.
17	Latch Block only.
98	Latch Shoe only.
96	Latch Rod Bracket only.
020	Draught Lever Bracket and Lever, complete with bolts. State, when ordering, if required for Home or Distant Signal.
0280	Balance Weight 56 lbs., complete with bolt.
030	Lever Number, complete with bolt and nut.

FITTINGS FOR MACHINE

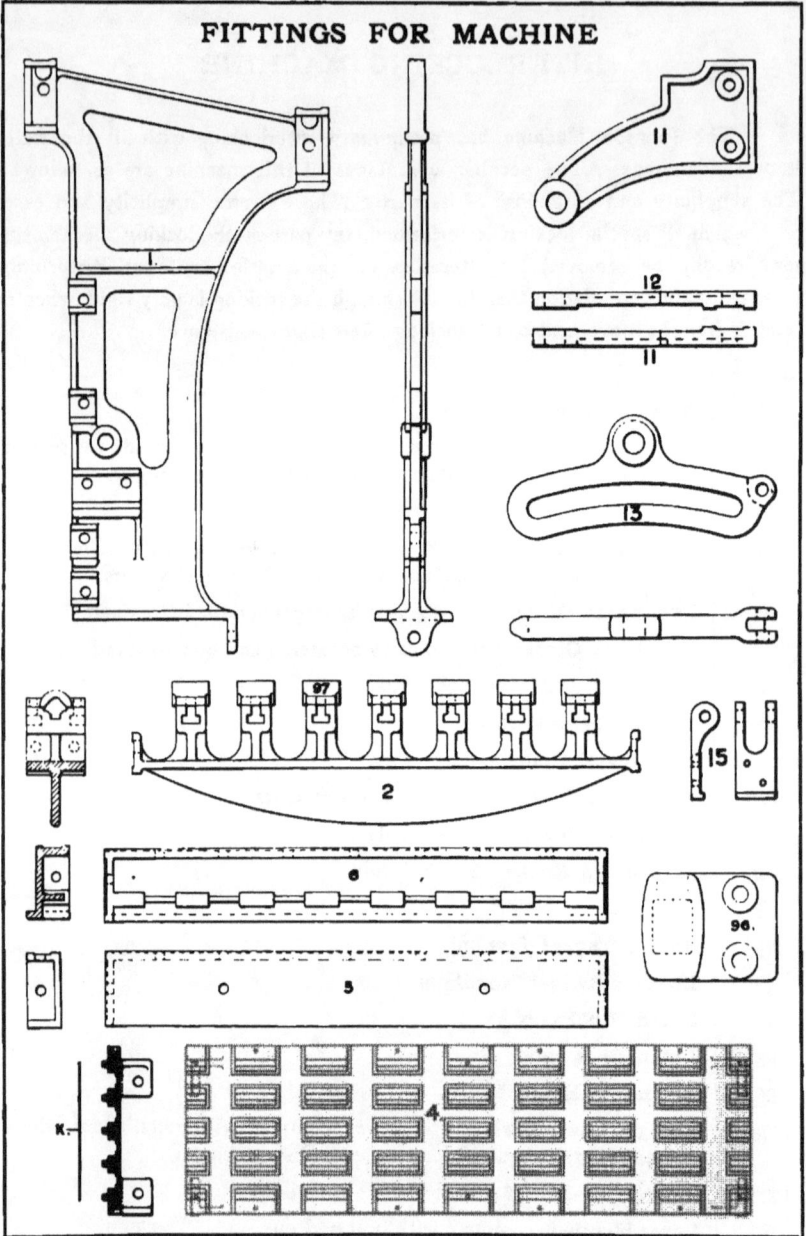

FITTINGS FOR MACHINE

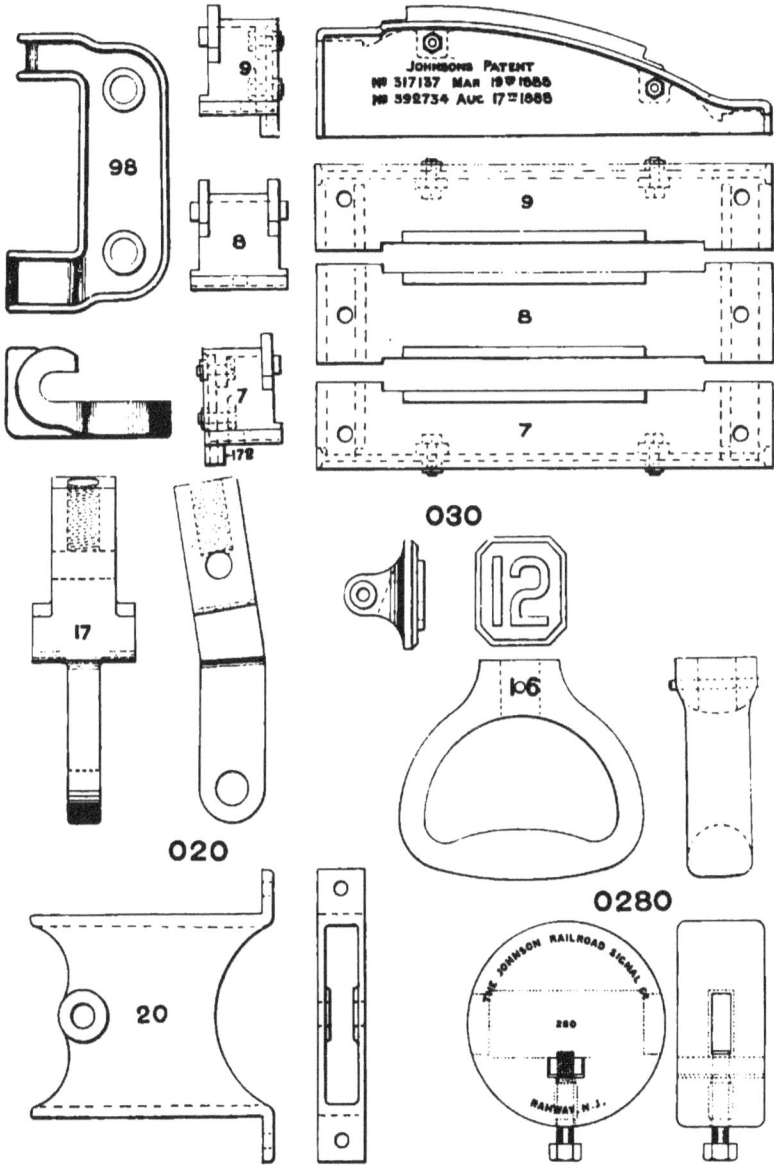

FITTINGS FOR MACHINE

FIG. 1

FIG. 2

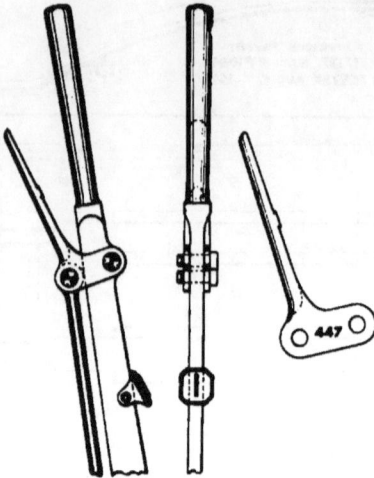

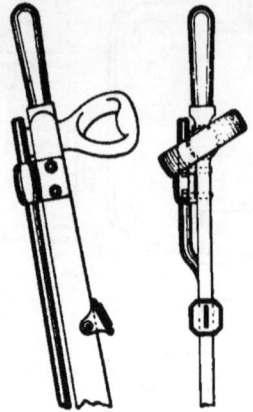

OH   OX

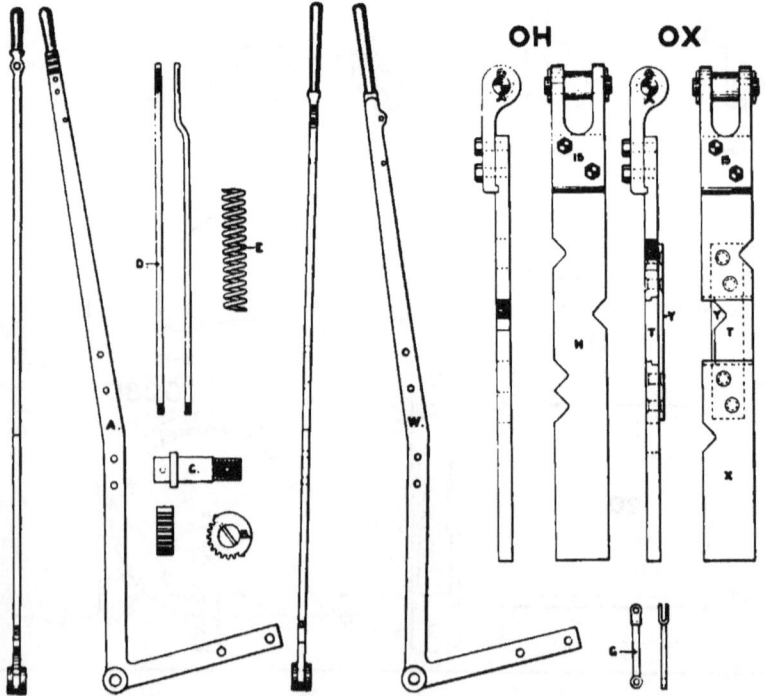

INTERLOCKING MACHINE—Continued

Order No.	
	Fig. **1**. Illustrates our Standard Lever, complete, with the ordinary latch handle.
	Fig. **2**. Illustrates Lever, complete, with swivel handle.
	Machine can be made with either kind of handle as desired.
OW	Standard Lever, complete, with rocker, rocker brackets, pins, cotters, latch block, shoe, latch rod, spring, latch handle, number and bolts.
W	Standard Lever only (Figure 1).
447	Latch Handle only, for standard lever.
AL	Latch Rod only, for standard lever.
OA	Lever, complete, with rocker, rocker brackets, pins, cotters, latch block, shoe, latch rod, spring, latch rod bracket, swivel handle, number and bolts.
A	Lever only (Figure 2).
B	Swivel Handle Pinion only.
C	Swivel Handle Pin only.
D	Latch Rod only.
E	Latch Spring only.
G	Rocker Coupling only.
OH	Ordinary Locking Tappet, complete, with jaw and pin.
H	Ordinary Locking Tappet only. In ordering give length.
J	Tappet Guide or Wrought-Iron Strips, $\frac{3}{8}$ inch x $\frac{1}{4}$ inch only. In ordering state number of levers. Illustrated on page 16.
K	Locking Plate Cover only. In ordering state number and depth of locking plate.
OX	Special Lock Tappet, complete, with jaw and pin.
X	Special Lock Tappet only.
Y	Special Lock Tappet Brace only.
Z	Draught Lever only, for Home Signal.
AB	Draught Lever only, for Distant Signal.
AC	Draught Lever Coupling only.

FITTINGS FOR MACHINE

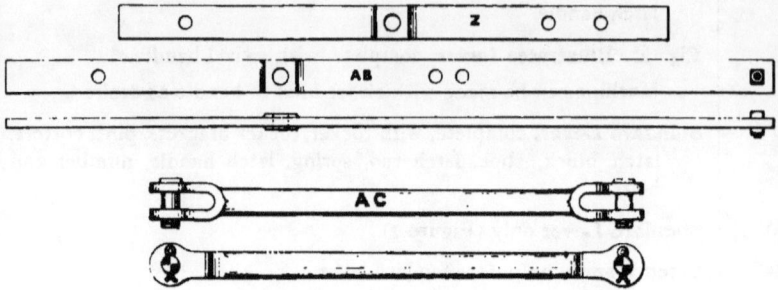

LOCKING DOGS AND TAPPETS

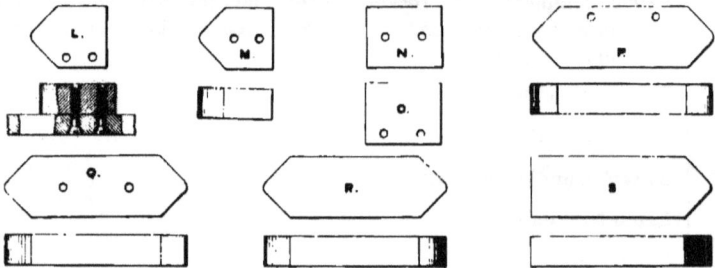

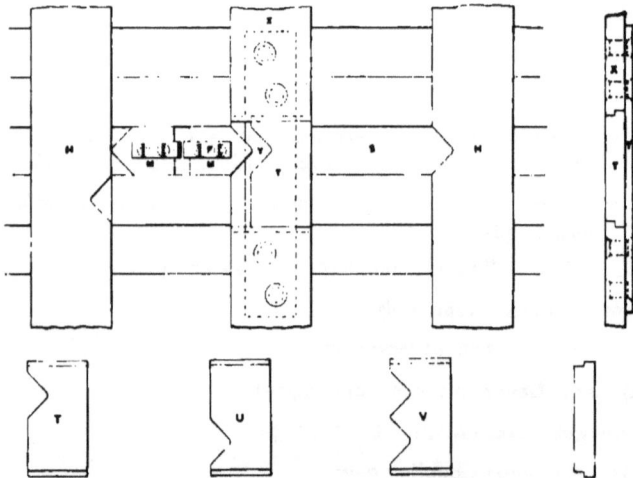

LOCKING DOGS AND TAPPETS

These locks are all made to standard gauges and are fixed to the locking bars by steel set screws.

Order No. |

L	No. 1. Locking Dog, with screws.
M	No. 2. Locking Dog, with screws.
N	No. 3. Locking Dog, with screws.
O	No. 4. Locking Dog, with screws.

No. 1 and 2 are the same, except the relative position of the screws. Nos. 3, 4, 5 and 6 are similar.

P	No. 5. Locking Dog, with screws.
Q	No. 6. Locking Dog, with screws.
R	No. 7. Locking Dog, with screws.
S	No. 8. Locking Dog, with screws.
T	No. 9. Sliding Tappet,
U	No. 10. Sliding Tappet,
V	No. 11. Sliding Tappet.
F	**Locking Bar**, $\frac{1}{8}$ inch x $\frac{1}{8}$ inch. In ordering state number of feet required.

THE JOHNSON
IMPROVED DWARF LEVER MACHINE

With or Without Locking

0247

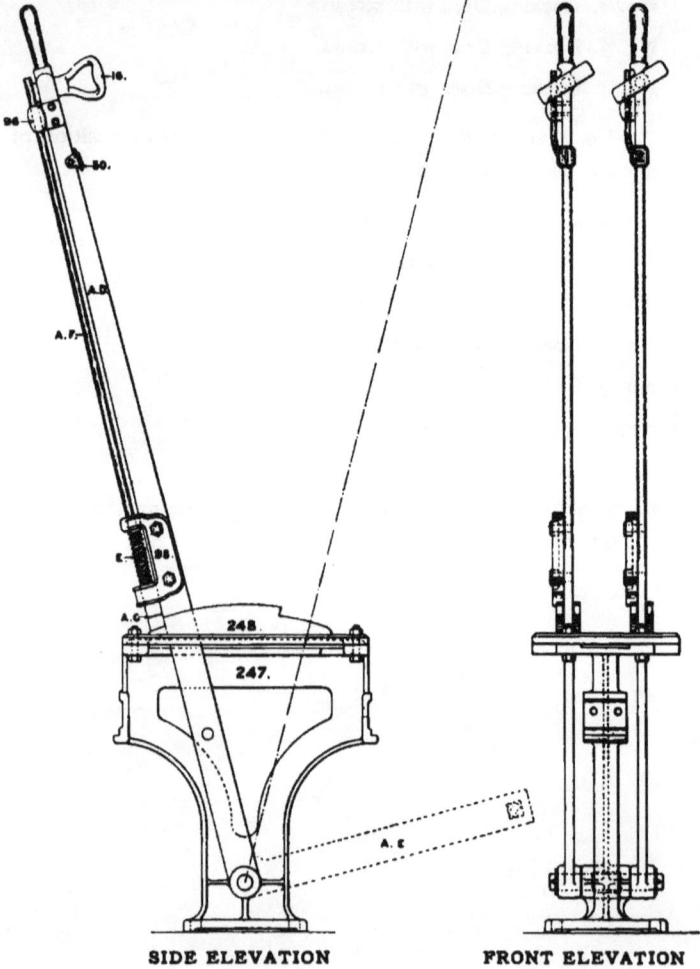

SIDE ELEVATION   FRONT ELEVATION

THE JOHNSON
IMPROVED DWARF LEVER MACHINE

With or Without Locking

This Apparatus was designed for use in simple cases, such for instance as Distant Switch Signals, or the protection of a siding connecting with the main line, etc. An elevated tower is often unnecessary at such places. At the same time any number of ground levers may be assembled where desirable and interlocked. The ground machine may be fixed in the open, as the locking, and other parts are designed for exposure to the weather. Either the straight or loop latch handle will be supplied with this machine, whichever is preferred.

Order No.	
0247	**Two Lever Machine**, complete, with stand, top plate and bolts, one switch lever complete, with loop handle, pinion, pin, latch rod block, spring, bracket, shoe, set screws, number and bolts, and one signal lever complete, with loop handle, pinion, pin, latch rod, block, spring, bracket, shoe, set screws, balance weight, number and bolts. In ordering state if locking is required.
247	**Stand**, for 2 levers only.
248	**Top Plate**, for 2 levers only.
AD	**Switch Lever** only.
AE	**Signal Lever** only.
AF	**Latch Rod** only.
AG	**Latch Block** only.
B	**Swivel Handle Pinion** only.
C	**Swivel Handle Pin** only.
E	**Latch Spring** only.
96	**Latch Rod Bracket** only.
98	**Latch Shoe** only.
280	**56 lbs. Balance Weight** only.
30	**Lever Number**, with bolt and nut.
16	**Loop or Swivel Handle** only.

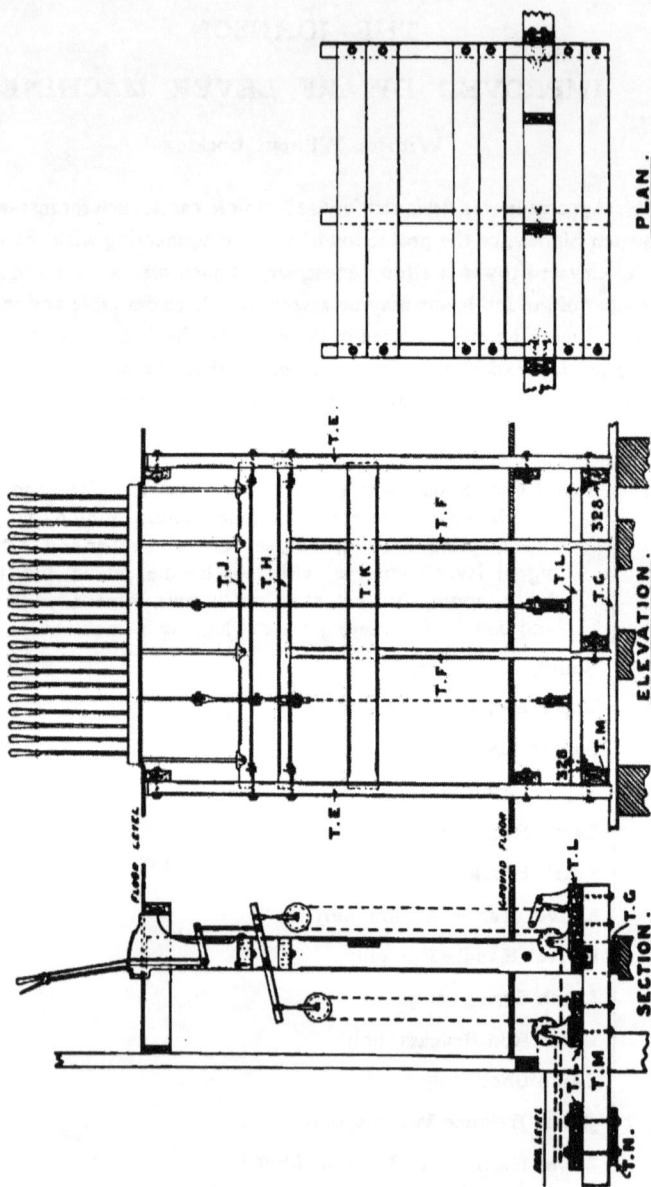

LEAD OUT AND FRAMING FOR 20 LEVER MACHINE

LEAD OUT

Lead Out and Framing for 20 Lever Machine

The object of this illustration is to show the framing of lead out timbers and machine supports, but in doing so we also give an outline of the machine and lead out to signals and switches. It will be seen that the upper sill on which the machine rests is supported by iron boxes resting on the intermediate sill, these boxes being also centre bearings, for the Gain Stroke Levers. When it so happens that gain stroke levers are not needed, as in the case when switches are worked direct from the tails of the levers, and cranks are used to lead out of the tower, then the iron boxes without the gain stroke levers are used, being placed directly under the legs of the machine. We beg to call attention to our improved arrangement of the lead out timbers. As will be seen from the illustration, the foundation for lead off cranks and wheels outside and directly in front of the tower, are by means of transverse timbers and bolts, tied to the timbers supporting the machine, and to the same timbers are bolted those which carry the cranks and wheels inside the tower.

FITTINGS

Order No.

TE	End Posts.
TF	Intermediate Posts.
TG	Bottom Sill.
TH	Intermediate Sill.
TJ	Top Sill.
TK	Rail, for pipe carriers.
TL	Bed Timbers, for lead out.
TM	Transverse Timbers.
TN	Bed Timber Sills.
328	Cast-Iron Brackets only.

When ordering any of these parts, give the size of the machine for which they are required including spare spaces. The Standard Tower is 13 feet from top of rail to floor level, and if the tower varies from this, the height should be stated when ordering the posts TE, TF.

CRANKS

O21

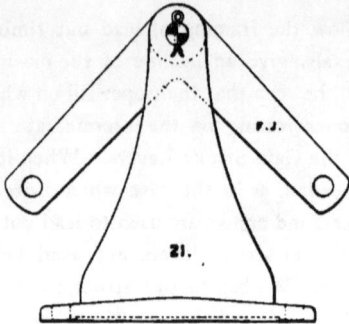 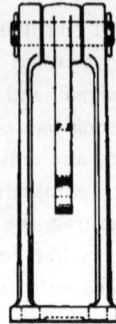

O83

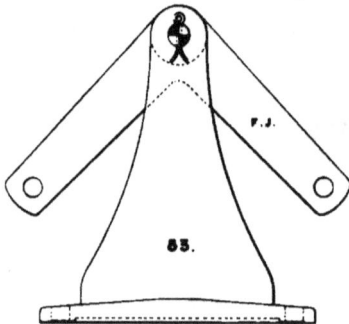 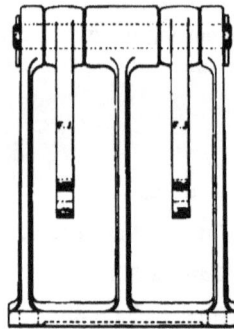

O84

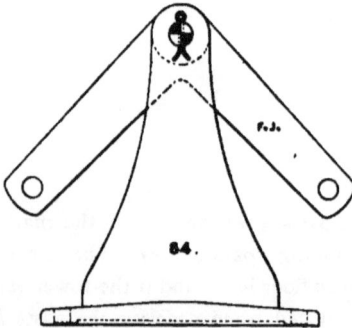 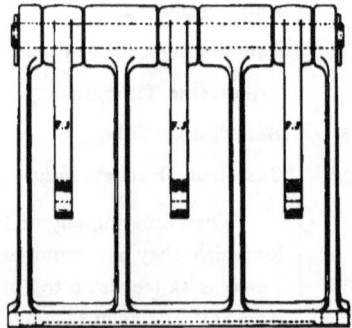

LEAD OUT—Continued

CRANKS

Order No.	
021	4-way Vertical Crank Stand, complete, with 9 inch x 9 inch wrought
F'J	9 inch x 9 inch Wrought Crank only.

BOX CRANKS

053	4-way Crank Box, complete with three 9 inch x 9 inch wrought hor. cranks, pins, cotters and bolts.
054	4-way Crank Box, complete, with four 9 inch x 9 inch wrought hor. cranks, pins, cotters and bolts.
056	6-way Crank Box, complete with five 9 inch x 9 inch wrought hor. cranks, pins, cotters and bolts.
057	6-way Crank Box, complete, with six 9 inch x 9 inch wrought hor. cranks, pins, cotters and bolts.
059	8-way Crank Box, complete, with seven 9 inch x 9 inch wrought hor. cranks, pins, cotters and bolts.
060	8-way Crank Box, complete, with eight 9 inch x 9 inch wrought hor. cranks, pins, cotters and bolts.
0287	10-way Crank Box, complete with nine 9 inch x 9 inch wrought hor. cranks, pins, cotters and bolts.
0288	10-way Crank Box, complete with ten 9 inch x 9 inch wrought hor. cranks, pins, cotters and bolts.

CRANKS
021

ERRATA, PAGE 33

ORDER NO. 021 *should read* 1-way Vertical Crank Stand, complete, with 9 inch x 9 inch wrought crank, pin, cotters and bolts, *and not* 4-way.

ORDER NO. 085 *should read* 4-way Vertical Crank Stand, complete, with four 9 inch x 9 inch wrought cranks, pin, cotters and bolts, *and not* 1-way.

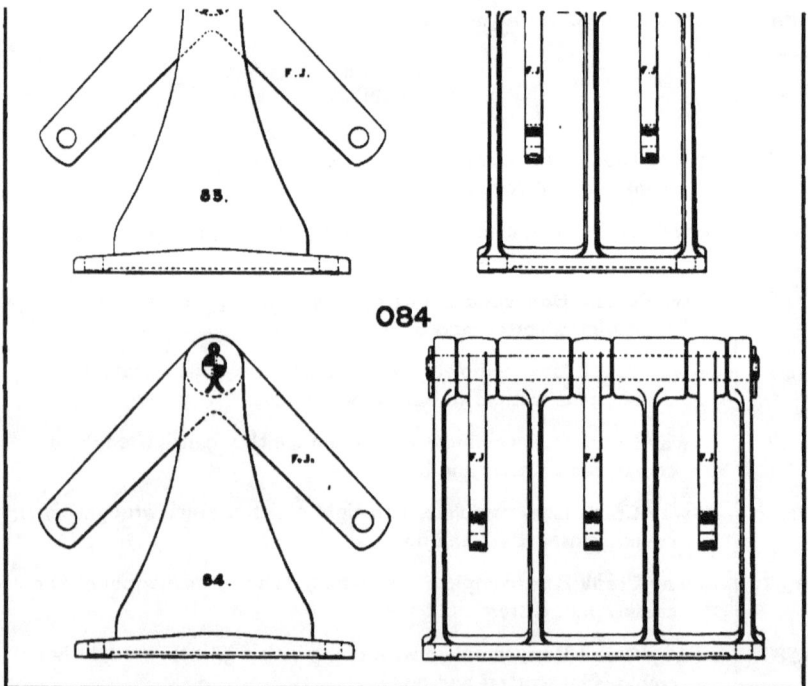

LEAD OUT—Continued

CRANKS

Order No.	
021	4-way Vertical Crank Stand, complete, with 9 inch x 9 inch wrought crank, pin, cotters and bolts.
083	2-way Vertical Crank Stand, complete, with two 9 inch x 9 inch wrought cranks, pin, cotters and bolts.
084	3-way Vertical Crank Stand, complete, with three 9 inch x 9 inch wrought cranks, pin, cotters and bolts.
085	1-way Vertical Crank Stand, complete, with four 9 inch x 9 inch wrought cranks, pin, cotters and bolts.
21	1-way Vertical Crank Stand only.
83	2-way Vertical Crank Stand only.
84	3-way Vertical Crank Stand only.
85	4-way Vertical Crank Stand only.
FJ	9 inch x 9 inch Wrought Crank only.

BOX CRANKS

053	4-way Crank Box, complete with three 9 inch x 9 inch wrought hor. cranks, pins, cotters and bolts.
054	4-way Crank Box, complete, with four 9 inch x 9 inch wrought hor. cranks, pins, cotters and bolts.
056	6-way Crank Box, complete with five 9 inch x 9 inch wrought hor. cranks, pins, cotters and bolts.
057	6-way Crank Box, complete, with six 9 inch x 9 inch wrought hor. cranks, pins, cotters and bolts.
059	8-way Crank Box, complete, with seven 9 inch x 9 inch wrought hor. cranks, pins, cotters and bolts.
060	8-way Crank Box, complete, with eight 9 inch x 9 inch wrought hor. cranks, pins, cotters and bolts.
0287	10-way Crank Box, complete with nine 9 inch x 9 inch wrought hor. cranks, pins, cotters and bolts.
0288	10-way Crank Box, complete with ten 9 inch x 9 inch wrought hor. cranks, pins, cotters and bolts.

CRANKS

085

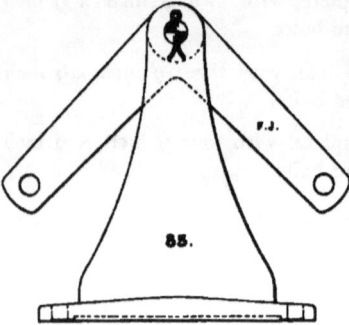
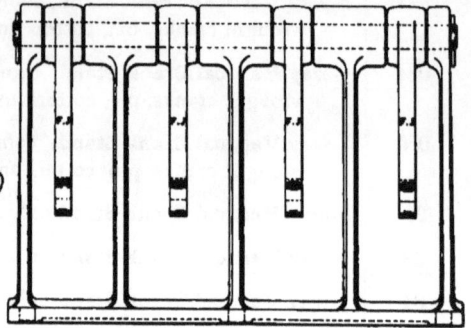

057

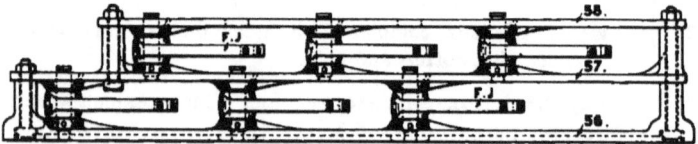

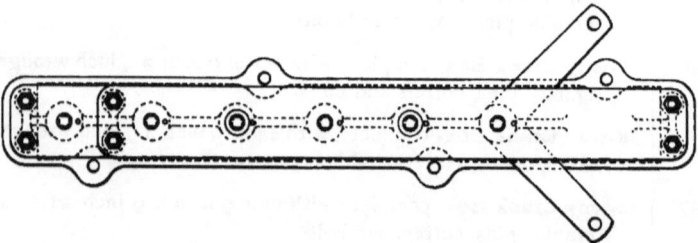

LEAD OUT—Continued

Order No.	
055	4-way Crank Box, with Bolts, Pins and Cotters only.
058	6-way Crank Box, with Bolts, Pins and Cotters only.
061	8-way Crank Box, with Bolts, Pins and Cotters only.
0280	10-way Crank Box, with Bolts, Pins and Cotters only.

It will be noticed that all our crank centre pins are supported at both ends, this allows us to use pins of 1 inch diameter instead of 1¼ inch as is usual, we thus increase the strength and reduce the friction. As far as possible we use wrought-iron cranks and jaws in preference to those made of malleable iron, although the latter are somewhat cheaper. Malleable work speedily becomes ramshackle and admits of lost motion.

P. R. R. STANDARD

AR	2¼ inch Rocking Shafts, left hand only.
AS	2¼ inch Rocking Shafts, right hand only.
307	Cap only, for 1, 2, 3 and 4-way bearings.
0300	1-way Rocking Shaft Bearings (in pairs), complete, with caps, set screws and bolts.
0302	2-way Rocking Shaft Bearings (in pairs), complete, with caps, set screws and bolts.
0304	3-way Rocking Shaft Bearings (in pairs), complete, with caps, set screws and bolts.
0306	4-way Rocking Shaft Bearings (in pairs), complete, with caps, set screws and bolts.
0400	1-way 2¼ inch Rocking Shafts, right hand, complete, with bearings, caps, set screws and bolts.
0401	2-way 2¼ inch Rocking Shafts, right hand, complete, with bearings, caps, set screws and bolts.
0402	3-way 2¼ inch Rocking Shafts, right hand, complete, with bearings, caps, set screws and bolts.
0403	4-way 2¼ inch Rocking Shafts, right hand, complete, with bearings, caps, set screws and bolts.

ROCKING SHAFTS

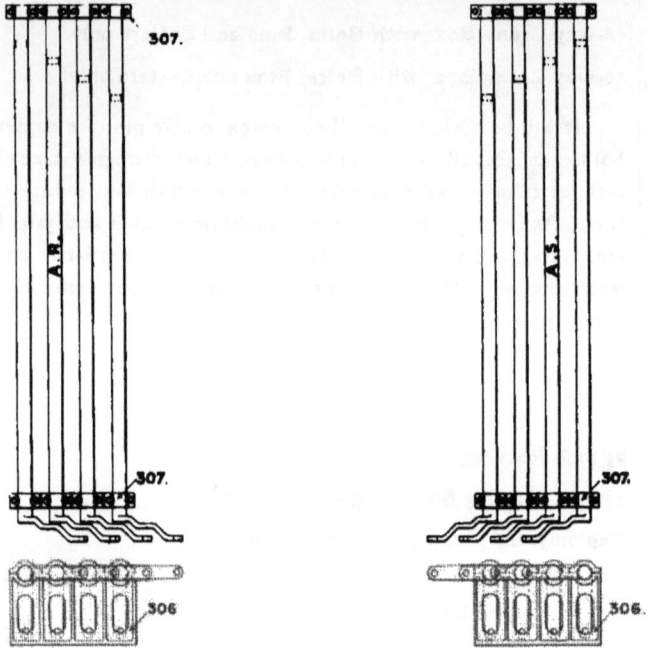

307.

A.R.

A.S.

307.

307.

306.

306.

LEAD OUT BRACKET

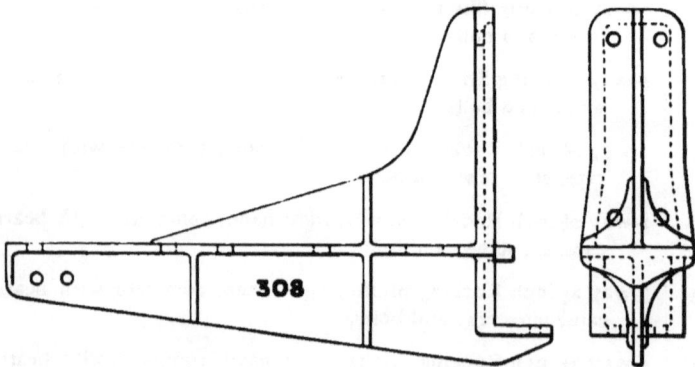

308

. LEAD OUT—Continued

Order No.

0404	1-way 2¼ inch **Rocking Shafts**, left hand, complete, with bearings, caps, set screws and bolts.
0405	2-way 2¼ inch **Rocking Shafts**, left hand, complete, with bearings, caps, set screws and bolts.
0406	3-way 2¼ inch **Rocking Shafts**, left hand, complete, with bearings, caps, set screws and bolts.
0407	4-way 2¼ inch **Rocking Shafts**, left hand, complete, with bearings, caps, set screws and bolts.
0308	**Lead Out Bracket**, right hand, complete, with bolts.
0309	**Lead Out Bracket**, left hand, complete, with bolts.
0310	**Lead Out Bracket**, intermediate, complete, with bolts.

WHEELS

0218	**No. 1. 1-way Vertical Wheel Stand**, complete, with 10 inch wheel, pin, cotters and lag screws.
0219	**No. 2. 1-way Vertical Wheel Stand**, complete, with 10 inch wheel, pin, cotters and lag screws.
0220	**No. 3. 1-way Vertical Wheel Stand**, complete, with 10 inch wheel, pin, cotters and lag screws.
0221	**No. 4. 1-way Vertical Wheel Stand**, complete, with 10 inch wheel pin, cotters and lag screws.
0222	**No. 1. 2-way Vertical Wheel Stand**, complete, with two 10 inch wheels, pin, cotters and lag screws.
0179	**No. 2. 2-way Vertical Wheel Stand**, complete, with two 10 inch wheels, pin, cotters and lag screws.
0223	**No. 3. 2-way Vertical Wheel Stand**, complete, with two 10 inch wheels, pin, cotters and lag screws.
0176	**No. 4. 2-way Vertical Wheel Stand**, complete, with two 10 inch wheels, pin, cotters and lag screws.
0224	**No. 1. 3-way Vertical Wheel Stand**, complete, with three 10 inch wheels, pin, cotters and lag screws.
0180	**No. 2. 3-way Vertical Wheel Stand**, complete, with three 10 inch wheels, pin, cotters and lag screws.

VERTICAL WHEELS

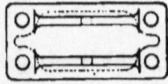
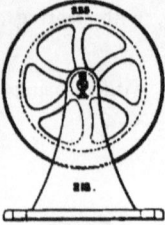

0218

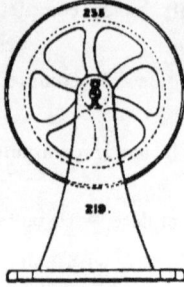

0219

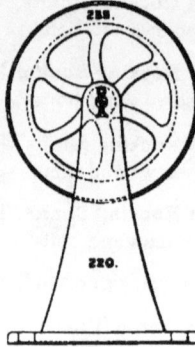

0220

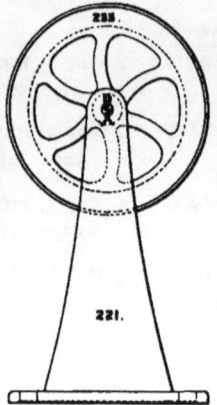

0221

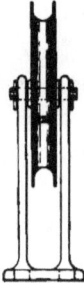
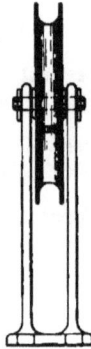
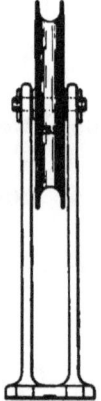

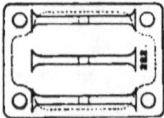
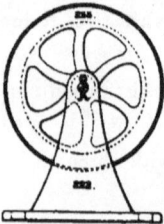

0222

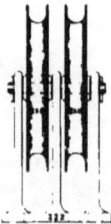
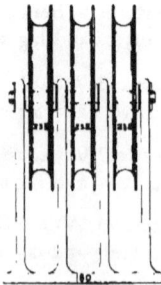

0180

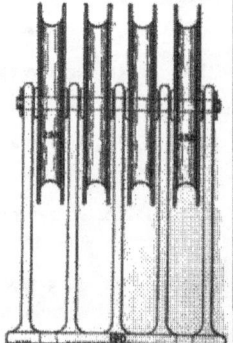

0190

LEAD OUT—Continued

Order No.	
0225	**No. 3. 3-way Vertical Wheel Stand**, complete, with three 10 inch wheels, pin, cotters and lag screws.
0177	**No. 4. 3-way Vertical Wheel Stand**, complete, with three 10 inch wheels, pin, cotters and lag screws.
0178	**No. 1. 4-way Vertical Wheel Stand**, complete, with four 10 inch wheels, pin, cotters and lag screws.
0181	**No. 2. 4-way Vertical Wheel Stand**, complete, with four 10 inch wheels, pin, cotters and lag screws.
0190	**No. 3. 4-way Vertical Wheel Stand**, complete, with four 10 inch wheels, pin, cotters and lag screws.
0191	**No. 4. 4-way Vertical Wheel Stand**, complete, with four 10 inch wheels, pin, cotters and lag screws.
218	**No. 1. 1-way Vertical Wheel Stand** only.
219	**No. 2. 1-way Vertical Wheel Stand** only.
220	**No. 3. 1-way Vertical Wheel Stand** only.
221	**No. 4. 1-way Vertical Wheel Stand** only.
222	**No. 1. 2-way Vertical Wheel Stand** only.
179	**No. 2. 2-way Vertical Wheel Stand** only.
223	**No. 3. 2-way Vertical Wheel Stand** only.
176	**No. 4. 2-way Vertical Wheel Stand** only.
224	**No. 1. 3-way Vertical Wheel Stand** only.
180	**No. 2. 3-way Vertical Wheel Stand** only.
225	**No. 3. 3-way Vertical Wheel Stand** only.
177	**No. 4. 3-way Vertical Wheel Stand** only.
178	**No. 1. 4-way Vertical Wheel Stand** only.
181	**No. 2. 4-way Vertical Wheel Stand** only.
190	**No. 3. 4-way Vertical Wheel Stand** only.
191	**No. 4. 4-way Vertical Wheel Stand** only.
234	6-inch **Wheel** only.
235	10-inch **Wheel** only.

FITTINGS FOR LEAD OUT

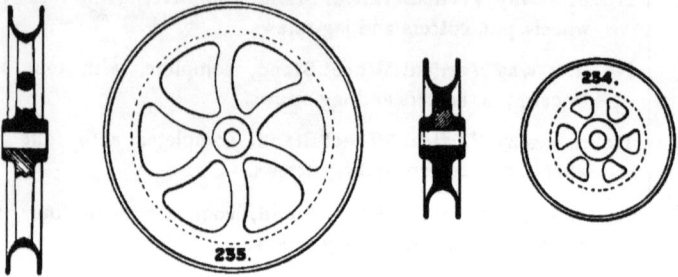

OCV

OAU

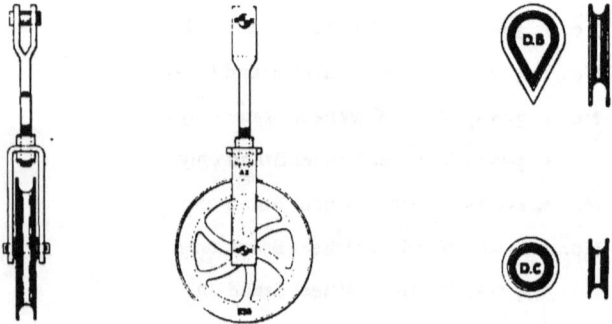

LEAD OUT—Continued

For convenience in leading out, we make four different heights of 1, 2, 3 and 4-way Vertical Wheels. The numbers 1 to 4 applied to each set of Vertical Wheel stands designate the height of the wheel from the foundation on which the stand is fixed, No. 1 being the lowest.

Order No.	
OAT	6 inch Draught Wheel, complete, with shackle, nuts, cotters and pins.
OAU	10 inch Draught Wheel, complete, with shackle, nuts, cotters and pins.
OAV	Shackle (consisting of jaw, fork and nuts), for 6 inch or 10 inch draught wheels.
AW	Jaw only.
AX	Fork only.
234	6 inch Wheel only.
235	10 inch Wheel only.
OCV	Wire Adjusting Screw, complete, with coupling, left and right hand screws and nuts.
CV	Coupling only.
CW	Left Hand Screw only.
CX	Right Hand Screw only.
CY	Chain only.
CZ	Ordinary Split Link only.
DB	Wire Eye, oval only.
DC	Wire Eye, round only.

PIPE CONNECTIONS

051	1-way Hor. Crank Stand, complete, with 8¾ inch x 9¼ inch wrought crank, pin, cotter and bolts.
052	2-way Hor. Crank Stand, complete, with one 8¾ inch x 9¼ inch and one 11½ inch x 12 inch wrought cranks, pin, cotter and bolts.

HORIZONTAL CRANKS

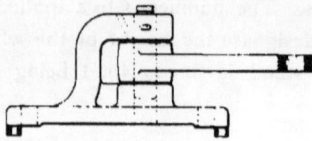

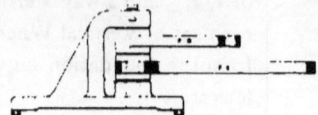

052

051

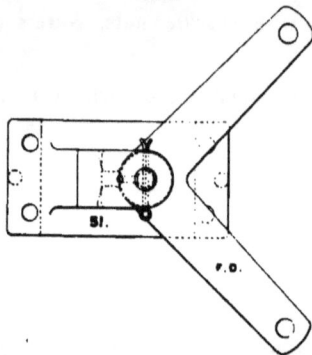

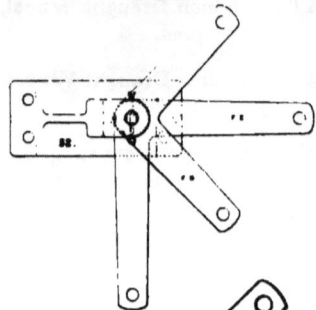

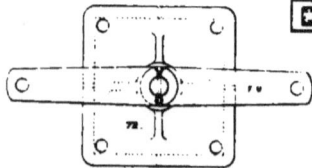

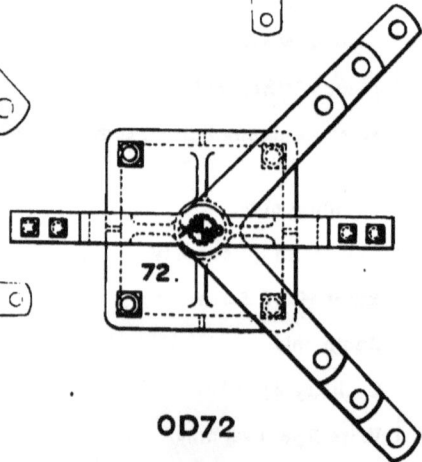

072   **0D72**

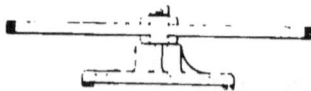

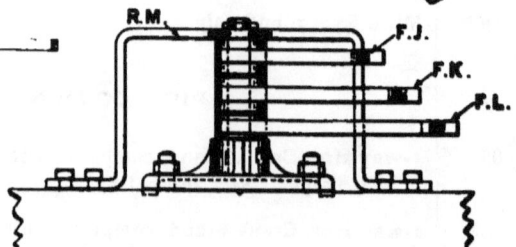

PIPE CONNECTIONS—Continued

Order No.	
O290	3-way Hor. Crank Stand, complete, with one 8¾ inch x 9¼ inch, one 11½ inch x 12 inch and one 14¼ inch x 14¾ inch wrought cranks, pin, cotter and bolts.
51	1-way Hor. Crank Stand only.
52	2-way Hor. Crank Stand only.
290	3-way Hor. Crank Stand only. (Not illustrated)
FD	8¾ inch x 9¼ inch, Wrought Crank only.
FE	11½ inch x 12 inch, Wrought Crank only.
FF	14¼ inch x 14¾ inch, Wrought Crank only.
0072	P. R. R. Standard 1-way Hor. Crank Stand, complete, with 9 inch x 9 inch wrought crank, pin, cotter and bolts.
OC72	P. R. R. Standard 2-way Hor. Crank Stand, complete, with one 9 inch x 9 inch, and one 11¾ inch x 11¾ inch wrought cranks, brace, pin, cotter and bolts.
OD72	P. R. R. Standard 3-way Hor. Crank Stand, complete with one 9 inch x 9 inch, one 11¾ inch x 11¾ inch and one 14½ inch x 14½ inch wrought cranks, brace, pin, cotter and bolts.
72	P. R. R. Standard 1-way Hor. Crank Stand only.
C72	P. R. R. Standard 2-way Hor. Crank Stand only.
D72	P. R. R. Standard 3-way Hor. Crank Stand only.
RL	Wrought Brace only, for 2-way Hor. Crank Stand.
RM	Wrought Brace only, for 3-way Hor. Crank Stand.
FJ	P. R. R. 9 inch x 9 inch Wrought Crank only.
FK	P. R. R. 11¾ inch x 11¾ inch Wrought Crank only.
FL	P. R. R. 14½ inch 14x ½ inch Wrought Crank only.
072	Hor. Compensator Stand, complete, with 18 inch x 18 inch wrought lever, pin, cotter and bolts.
OA72	Hor. Compensator Stand, complete, with 21 inch x 21 inch wrought lever, pin, cotter, and bolts.
OB72	Hor. Compensator Stand, complete with 24 inch x 24 inch, wrought lever, pin, cotter and bolts.

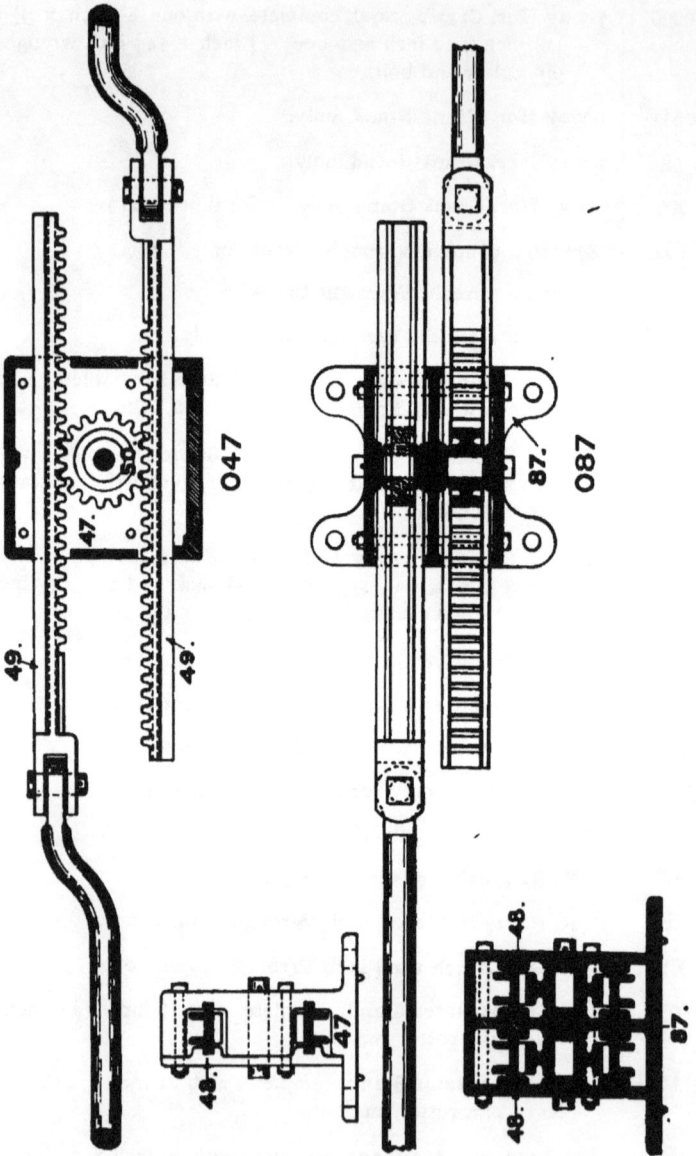

THE JOHNSON RACK AND PINION COMPENSATOR

Pat. No. 309627, Dec. 23, 1884

PIPE CONNECTIONS—Continued

Order No.	
72	Horizontal Compensator Stand only.
FU	18 inch x 18 inch Wrought Compensator Lever only.
FV	21 inch x 21 inch Wrought Compensator Lever only.
FW	24 inch x 24 inch Wrought Compensator Lever only.
0297	1-way Vertical Compensator Stand, complete, with one 9 inch x 9 inch wrought lever, pin, cotters and bolts.
0298	2-way Vertical Compensator Stand, complete, with two 9 inch x 9 inch wrought levers, pin, cotters and bolts.
0299	3-way Vertical Compensator Stand, complete, with three 9 inch x 9 inch wrought levers, pin, cotters and bolts.
297	1-way Vertical Compensator Stand only.
298	2-way Vertical Compensator Stand only.
299	3-way Vertical Compensator Stand only.

IMPROVED RACK AND PINION COMPENSATOR

We invite special attention to our Improved Rack and Pinion Compensator. The diameter of the pinion has been increased. The racks are strengthened and guided, by planed flanges, sliding through planed grooves in the box. The box is perforated, so as to facilitate the oiling of the pinion and racks.

The advantages of a rack and pinion compensator are briefly as follows. It cannot be dead centred, by a long stroke, and great variation in temperature. This is not true of other compensators. One compensator will answer for any length of rod. It is very compact and does not come in the way of other connections.

047	1-way Rack and Pinion Compensator, complete, with stand, two racks, pinion, side plate, pins, cotters and bolts.
087	2-way Rack and Pinion Compensator, complete, with stand, four racks, two pinions, two side plates, pins, cotters and bolts.
47	1-way Rack and Pinion Compensator Stand only.
87	2-way Rack and Pinion Compensator Stand only.

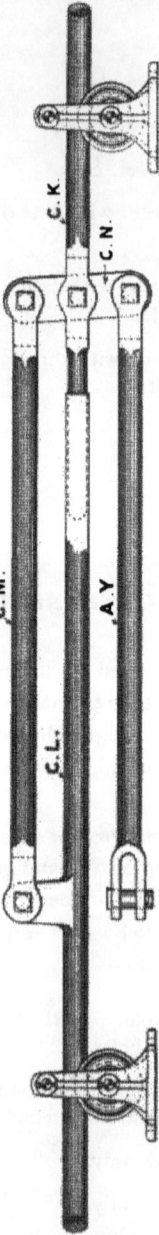

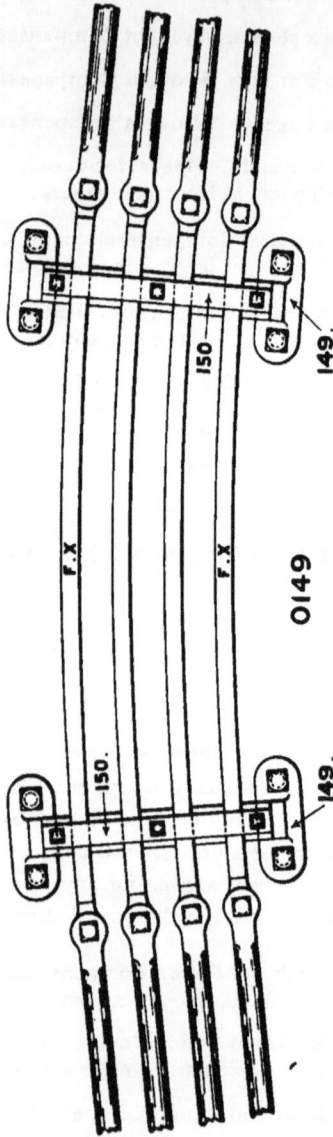

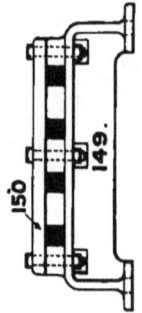

THE JOHNSON EQUALIZER

(Patented)

PIPE CONNECTIONS—Continued

Order No.	
49	**Rack** only.
50	**Pinion** only.
48	**Side Plate** only, for stand.

EQUALIZER

The equalizing lever supplies a long felt want, in working a crossover road, or more than one switch by one lever.

The difficulty has been that it was almost impossible to adjust and keep adjusted the connections, and stroke, so as to make both switches work equally well up to the stock rail. This was caused by one switch checking the complete action of the other. By the use of an Equalizer, equal pressure is exerted on both switches, and it is made impossible for one switch to check the other.

OCK	**Equalizing Lever,** complete, with double jaw, lug, coupling, 5½ inch lever, two 1¼ inch jaws, cotters and pins.
CK	**Equalizing Lever, Double Jaw** only.
CL	" " **Lug** only.
CM	" " **Coupling** only.
CN	" " **5½ inch Lever** only.
AY	" " **Two 1¼ inch Jaws** only.

ANGLE BARS

We were the first to introduce the angle bar, and it has been more successful than any other device for easing off curves.

0143	**1-way Angle Bar Stand,** complete, with bar, two stands, caps, bolts and lag screws.
0145	**2-way Angle Bar Stand,** complete, with two bars, two stands, caps, bolts and lag screws.
0147	**3-way Angle Bar Stand,** complete, with three bars, two stands, caps, bolts and lag screws

JAWS

OAY

OBC

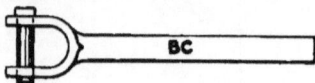

OAZ

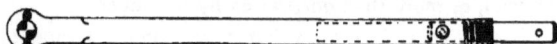

OBA

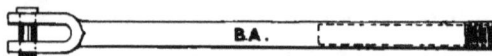

OBB

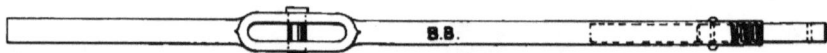

OBD

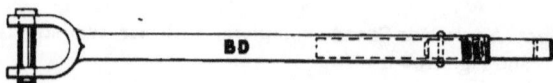

PIPE CONNECTIONS—Continued

Order No.	
0149	4-way Angle Bar Stand, complete, with four bars, two stands, caps, bolts and lag screw.
143	1-way Angle Bar Stand only.
144	1-way Angle Bar Stand Cap only.
145	2-way Angle Bar Stand only.
146	2-way Angle Bar Stand Cap only.
FX	Angle Bar only.

These Angle Bar Stands are made from 1-way up to any number required. In ordering specify the number of bars. The illustration shows a 4-way stand.

JAWS

OAY	1¼ inch Wrought Solid Jaw, complete, with pin and cotter.
OAZ	1¼ inch Wrought Jaw, complete, with plug, pipe screwed end, pin and cotter.
OBA	1¼ inch Wrought Jaw, complete, with pipe screwed end, pin and cotter.
OBB	1¼ inch Wrought Double Jaw, complete, with one end solid, the other end with plug, pipe screwed end, pin and cotter.
OBC	1¼ inch Wrought Solid Wide Jaw, complete, with pin and cotter.
OBD	1¼ inch Wrought Wide Jaw, complete, with plug, pipe screwed end, pin and cotter.
OBE	1¼ inch Wrought Solid Jaw, complete, with plug and pipe screwed end.
OBF	1¼ inch Wrought Lug, complete, with one end solid the other end with plug and pipe screwed end.
OFY	1¼ inch Screw Jaw, complete, with malleable jaw, solid screwed end and nut, pin and cotter.
271	Malleable Jaw only.
OFZ	1¼ inch Screw Jaw, complete, with malleable jaw, solid screwed end and nut, plug and pipe screwed end, pin and cotter.

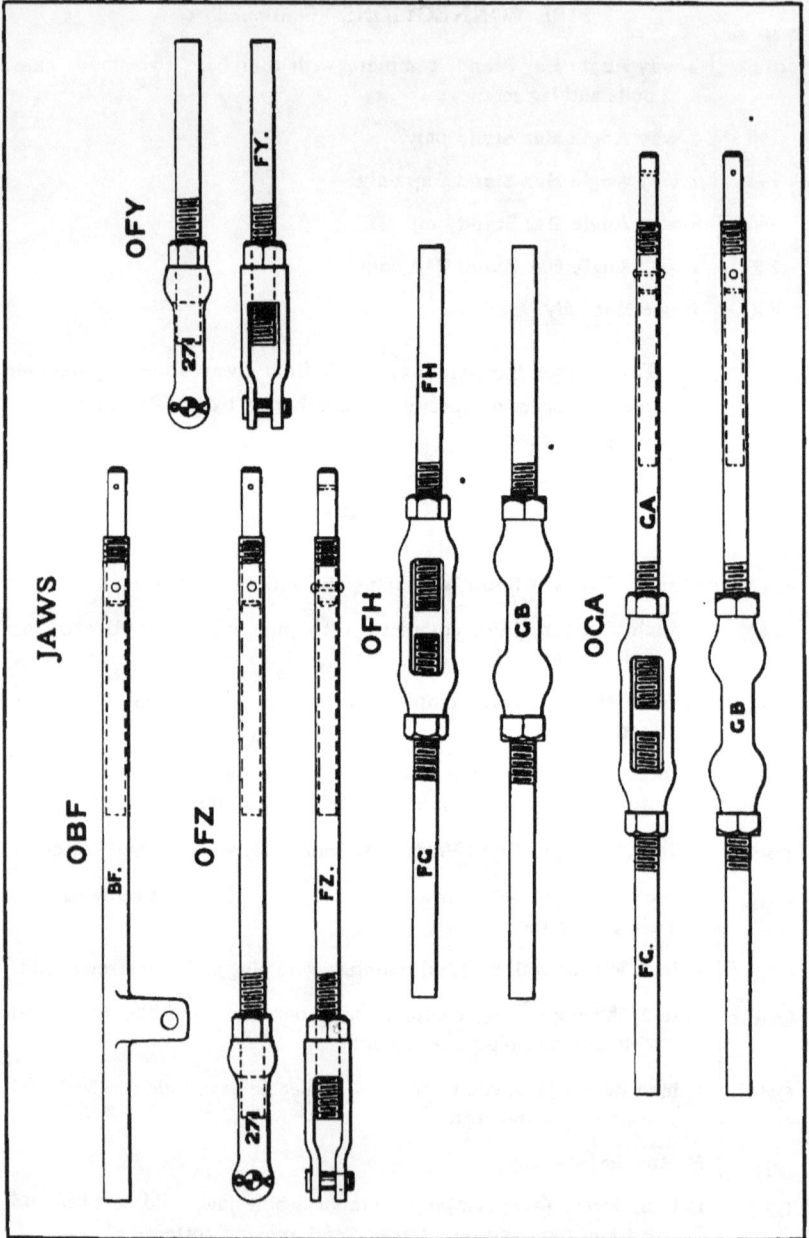

PIPE CONNECTIONS—Continued

Order No.	
OFH	**Pipe Adjusting Screw,** complete, with socket, left and right solid screwed ends and nuts.
GB	**Wrought Socket** only.
OGA	**Pipe Adjusting Screw,** complete, with socket, one solid screwed end, the other end with plug and pipe screwed end and nuts.
OGC	**Joint, for 1 inch Pipe,** complete, with sheve, plug and 2 rivets.
GC	**Sheve, for 1 inch Pipe** only.
GD	**Plug, for 1 inch Pipe** only.
GE	**Rivets, for 1 inch Pipe** only.
GF	**1 inch Pipe,** in lengths of 16 feet, 17 feet and 18 feet, one end screwed, the other end with sheve and plug.
064	**Style A, 1-way Pipe Carrier,** complete, with stand, sheve, roller, pins, cotters and lag screws.
065	**Style A, 2-way Pipe Carrier,** complete, with stand, two sheves, two rollers, pins, cotters and lag screws.
0064	**Style A, 3-way Pipe Carrier,** complete, with two 1-way stands, three sheves, three rollers, pin, cotters and lag screws.
0065	**Style A, 4-way Pipe Carrier,** complete, with one 1-way stand and one 2-way stand, four sheves, four rollers, pins, cotters and lag screws.
64	**Style A, 1-way Stand** only.
65	**Style A, 2-way Stand** only.
66	**3 inch Sheve** only.
66A	**Top Roller** only.

These Pipe Carriers are made from 1-way up to any number required. In ordering state number of ways.

ANTI-FRICTION PIPE CARRIER

The Anti-Friction Pipe Carrier increases the power possible to be exerted on switches, and facing point locks, because the friction on the working connection is reduced to a minimum. The Anti-Friction Carrier shown in the illustration (page 52), was designed for use both on curved and straight connections.

PIPE CARRIERS

064

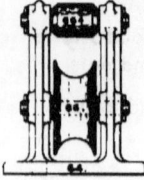

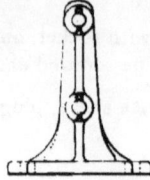

0065

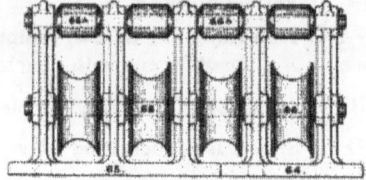

065

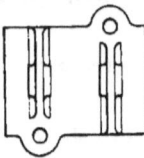

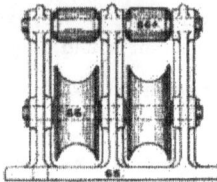

0064

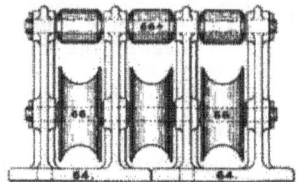

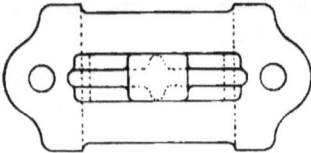

0236

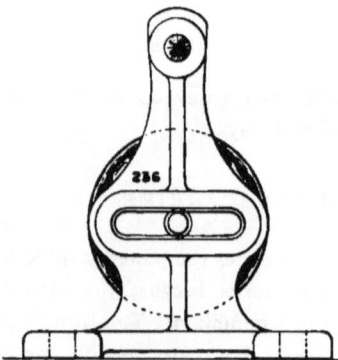

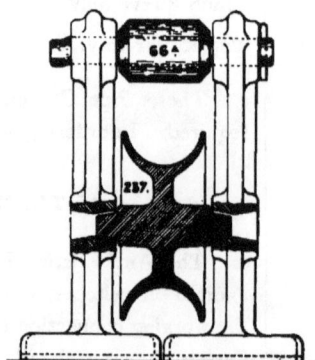

PIPE CONNECTIONS—Continued

Order No.

0236   Style B, Anti-Friction Pipe Carrier, complete, with stand, sheve, roller, pin, cotter and lag screws.

236   Style B, Anti-Friction Pipe Carrier Stand only

237   Style B, 3 inch Sheve only.

66A   Style B, Top Roller only, the same as Style A Pipe Carrier

These Pipe Carriers are made from 1-way up to any number required. In ordering state number of ways.

SELECTORS

We are the original inventors of the system of the controlling a number of Signals by one lever and selecting the signal to be given, by the position of the switches. Our selector, which is widely used, will be found to be specially adaptable to the selection of three or more signals. Selectors may be fixed vertically in the Tower or alongside the ground connections, whichever suits the case most economically.

043   2-way Selector (to work 2 blades), pipe only, complete, with selector stand, cover, bearings, caps and bolts, slide with pin and cotters, eye joint, two connecting hooks, two cams, collar, crank and set screws, shaft, double jaw with pin and cotter, two slide bars, with bolts, pins and cotters, guide stand and cover with bolts and lag screws, two 1¼ inch jaws with pins and cotters, 2-way roller carrier stand with two rollers, pin, cotters and lag screws.

0293   3-way Selector (to work 3 blades), pipe only, complete, with selector stand, cover, bearings, caps and bolts, slide with pin and cotters, eye joint, three connecting hooks, three cams, four collars, two cranks and set screws, two shafts, two double jaws with pins and cotters, three slide bars with bolts, pins and cotters, guide stand and cover with bolts and lag screws, three 1¼ inch jaws with pins and cotters, 3-way roller carrier stand, with three rollers, pin, cotters and lag screws

033   4-way Selector (to work 4 blades), pipe only, complete with selector stand, cover, bearings, caps and bolts, slide with pin and cotters, eye joint, four connecting hooks, eight cams, six collars, three

THE JOHNSON PATENT SELECTOR

Pat. No. 343911, June 1st, 1886

O43

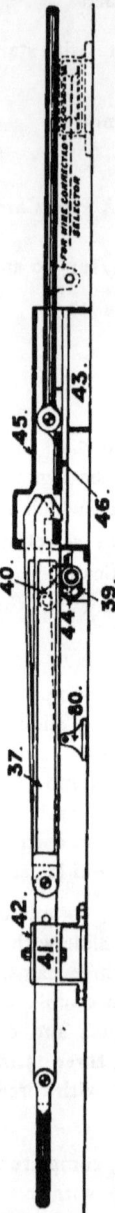
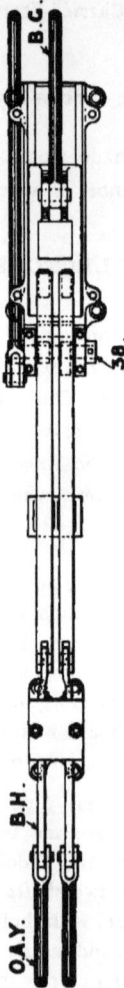
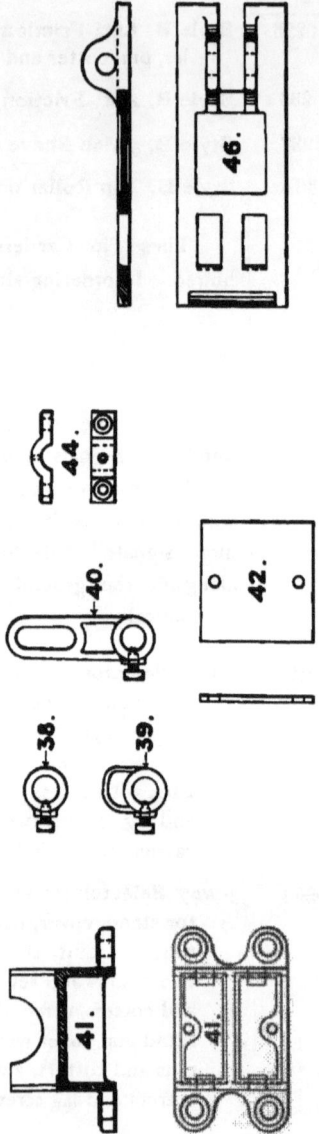

FITTINGS

PIPE CONNECTIONS—Continued

cranks and set screws, three shafts, three double jaws with pins and cotters, four slide bars with bolts, pins and cotters, guide stand and cover with bolts and lag screws, four 1¼ inch jaws with pins and cotters, 4-way roller carrier stand, with four rollers, pin, cotters and lag screws.

0043 **2-way Selector, Pipe and Wire,** complete, with selector stand, cover, bearings, caps and bolts, slide with pin and cotters, eye joint with lug, two connecting hooks, two cams, collar, crank and set screws, shaft, double jaw with pin and cotter, two slide bars with bolts, pins and cotters, guide stand and cover with bolts and lag screws, 2-way roller carrier stand with two rollers, pin, cotters and lag screws, 1-way 6 inch hor. wheel stand, with 6 inch wheel, shackles, pins, wire eyes and lag screws.

00293 **3-way Selector, Pipe and Wire,** complete, with selector stand, cover, bearings, caps and bolts, slide with pin and cotters, eye joint with lug, three connecting hooks, three cams, four collars, two cranks and set screws, two shafts, two double jaws with pins and cotters, three slide bars with bolts, pins and cotters, guide stand and cover with bolts and lag screws, 3-way roller carrier stand with three rollers, pin, cotters and lag screws, 1-way 6 inch hor. wheel stand with 6 inch wheel, shackles, pins, wire eyes and lag screws.

0033 **4-way Selector, Pipe and Wire,** complete, with selector stand, cover, bearings, caps and bolts, slide with pin and cotters, eye joint with lug, four connecting hooks, eight cams, six collars; three cranks and set screws, three shafts, three double jaws with pins and cotters, four slide bars with bolts, pins and cotters, guide stand and cover with bolts and lag screws, 4-way roller carrier stand, with four rollers, pin, cotters and lag screws, 1-way 6 inch hor. wheel stand, with 6 inch wheel, shackles, pins, wire eyes and lag screws.

FITTINGS FOR SELECTORS

41 2-way Guide Stand only.

291 3-way Guide Stand only.

31 4-way Guide Stand only.

42 2-way Guide Stand Cover only.

292 3-way Guide Stand Cover only.

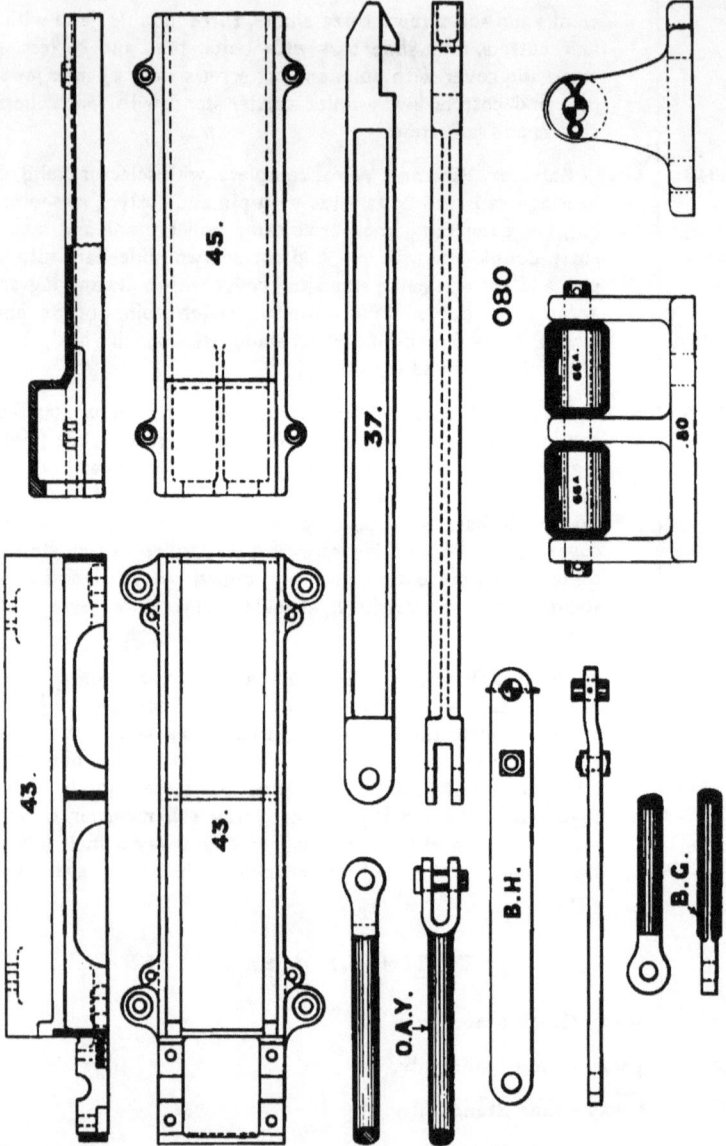

FITTINGS FOR SELECTORS

PIPE CONNECTIONS—Continued

Order No.	
32	4-way Guide Stand Cover only.
43	2-way Selector Stand only.
293	3-way Selector Stand only.
33	4-way Selector Stand only.
45	2-way Selector Stand Cover only.
295	3-way Selector Stand Cover only.
35	4-way Selector Stand Cover only.
44	2-way Bearing Cap only.
294	3-way Bearing Cap only.
34	4-way Bearing Cap only.
46	2-way Slide only.
296	3-way Slide only.
36	4-way Slide only.
37	Connecting Hook only.
38	Collar and Set Screw only.
39	Cam and Set Screw only.
40	Crank Arm and Set Screw only.
BG	Eye Joint only.
SV	Eye Joint with Lug only.
BH	2-way Slide Bars only.
ST	3-way Slide Bars only.
SU	4-way Slide Bars only.
080	Roller Carrier for 2-way Selector, complete, with stand, two rollers, pin, cotters and lag screws.
081	Roller Carrier for 3-way Selector, complete, with stand, three rollers, pin, cotters and lag screws.
082	Roller Carrier for 4-way Selector, complete, with stand, four rollers, pin, cotters and lag screws.
0214	6 inch Hor. Wheel Stand, complete, with 6 inch wheel, pin, cotters and lag screws.

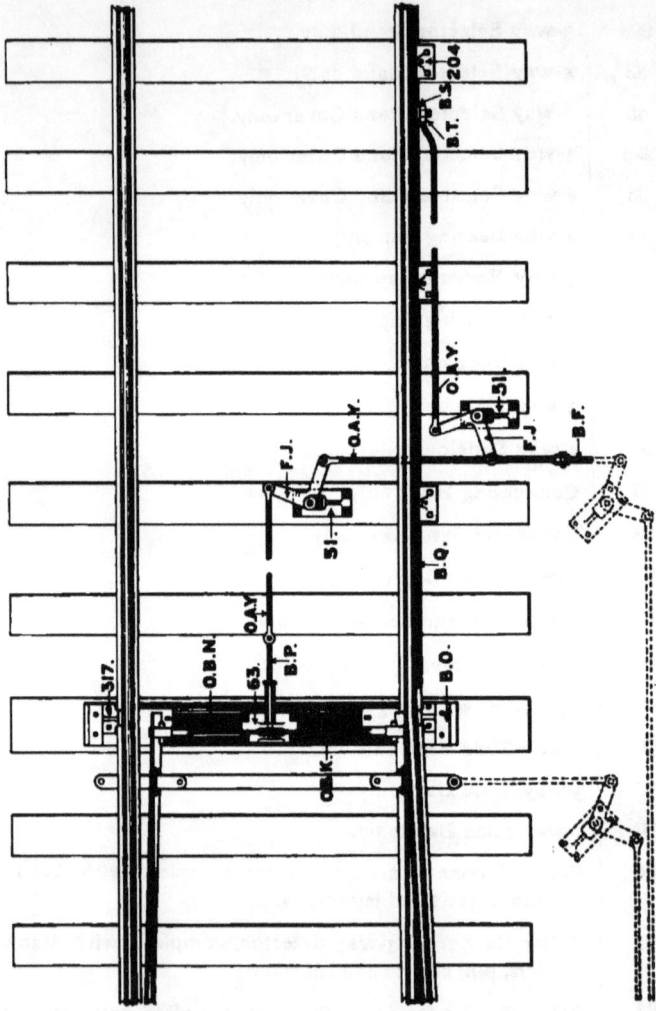

FACING POINT LOCK
STYLE A
063

Order No.	
OAY	1¼ inch **Wrought Jaw**, complete, with pin and cotter.
OBB	1¼ inch **Double Jaw**, complete, with pin and cotter.
SW	1¼ inch **Solid Shaft** only, 24 inches long.
CT	⅞ inch **Shackle** only.
DB	**Wire Eye**, oval only.

FACING POINT LOCKS

The special feature about our Facing Point Locks is the Detector Bar, shown in the illustration (page 60). This bar is very strong and simple, and works very easily, and is made in two lengths with a coupling bar, this being more easy to handle and fix on the rails. Our invention and improvement consist in beveling the bar and inclining it toward the rail, to prevent it sagging away from the rail; also in the cam movement for raising the bar. Our detector bar will not become clogged by snow or anything else.

When ordering send section of rail.

Order No.	
063	**Style A, Facing Point Lock**, complete, with 40 feet Detector Bar, consisting of two 18 feet bars with 5 feet 6 inch coupling bar, ten rail clips with rollers, pins, cotters and ten pairs of Bush interlocking bolts, ten slide plates with studs and rivets, driving plate with pin, cotter, washer and rivets, eye joint, tie bar 38 feet long with spikes, six 1¼ inch jaws with pins and cotters, two 9 inch x 9 inch wrought cranks and stands with pins, cotters, bolts and washers, wrought lug, plunger stand with round steel plunger, roller, bolts, pin, washers and cotters, front rod (Style C, unless otherwise ordered) with lugs, bolts, pins and cotters, iron plate with one end strip rivetted on and one loose, two rail braces.
0063	**Style A, Facing Point Lock** only, complete, with plunger stand with round steel plunger, roller, bolts, pin, washers and cotters, front rod (Style C, unless otherwise ordered) with lugs, bolts, pins and cotters, iron plate with one end strip rivetted on and one loose, two rail braces, 1¼ inch jaw with pin and cotter.

FITTINGS FOR STYLE A, FACING POINT LOCK

Order No.	
OA63	**Plunger Stand**, complete, with round steel plunger, roller, bolts, pin, washer and cotters.

DETECTOR BAR FOR SWITCH AND LOCK MOVEMENT

Pat. No. 377430, Feb. 7, 1888

STYLE A
0204

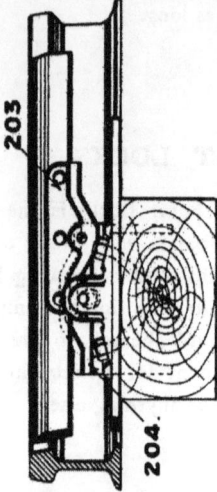
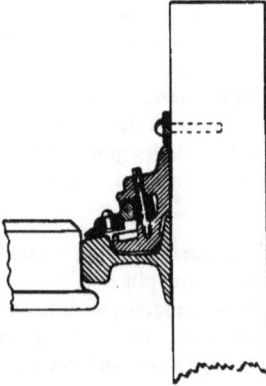

STYLE B
0243

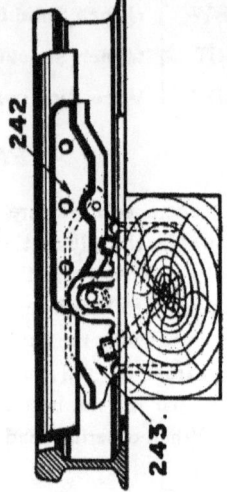
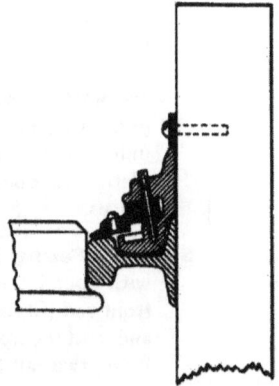

FACING POINT LOCKS—Continued

Order No.	
63	Plunger Stand only.
73	Plunger Stand Roller only.
BP	Round Steel Plunger only.
0204	Detector Bar (Style A), complete, with two 18 feet bars with 5 feet 6 inch coupling bar, ten rail clips with rollers, pins, cotters and ten pairs of Bush interlocking bolts, ten slide plates with studs and rivets, driving plate with pin, cotter, washer and rivets, eye joint, tie bar 38 feet long with spikes.
203	Slide Plate only.
204	Rail Clip only. When ordering send section of rail.
BX	Bush Interlocking Bolt only.
BQ	Detector Bar only.
BR	Coupling, only for Detector Bar.
BS	Driving Plate only for Detector Bar.
BT	Eye Joint only for Detector Bar.
BU	Tie Bar only.
BV	3 Arm Wrought Crank only.
OBK	Front Rod (Style A), complete, with pins and cotters.
BK	Front Rod only.
BL	Pin only.
BM	Cotters only.
OSR	Front Rod (Style C), complete, with two lugs, pins, cotters and bolts.
SR	Front Rod only.
SS	Lugs only.
OBN	Iron Plate, complete, with one end strip rivetted on and one loose.
OAY	1¼ inch Wrought Jaw, complete, with pin and cotter.
BF	1¼ inch Wrought Lug only.
051	1-way Hor. Crank Stand, complete, with 9 inch x 9 inch wrought crank, pin, cotter, bolts and washers.
317	Rail Braces only.

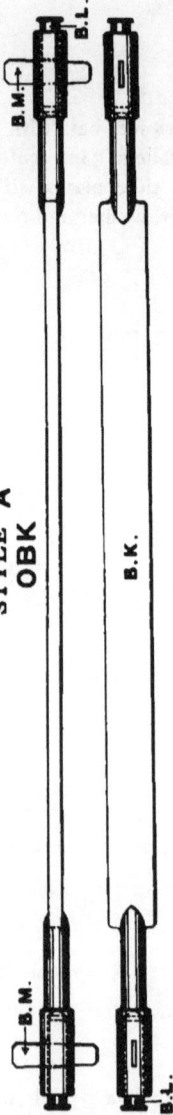

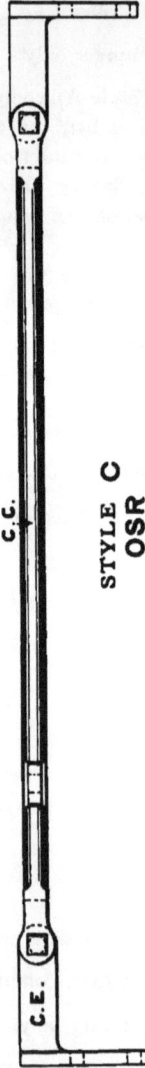
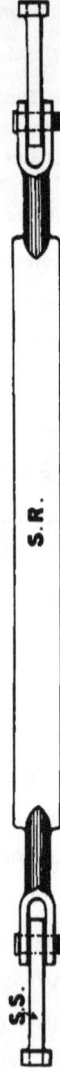
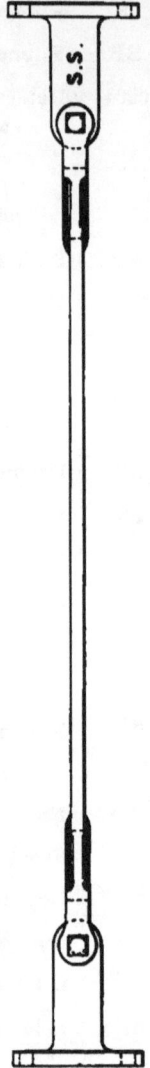

FRONT RODS FOR FACING POINT LOCKS

STYLE A
OBK

STYLE B
OCC

STYLE C
OSR

FACING POINT LOCKS—Continued

SWITCH AND LOCK MOVEMENT

In all good switch and lock movements the Detector Bar is completely raised before any movement of the switch takes place. This necessitates providing an active and silent movement to drive the bar as well as the switch. In our Switch and Lock Movement the Detector Bar makes its own silent action as regards vertical movement, and as we do not check the action of the bar when once it is started, its impetus helps to carry over the switch. We draw special attention to our Adjustable Disc for regulating the stroke of the switch. Practical tests have shown that this Switch and Lock Movement works very easily at a distance of 1,100 feet from the Signal Tower.

Order No.	
0244	**Switch and Lock Movement**, complete, with 40 feet Detector Bar (Style B), consisting of two 18 feet bars with 5 feet 6 inch coupling bars, ten rail clips with rollers, pins, cotters and ten pairs of Bush interlocking bolts, ten slide plates with studs and rivets, driving plate with pin, cotter, washer and rivets, eye joint, tie-bar 38 feet long with spikes, two 1¼ inch wrought jaws with pins and cotters, two 1¼ inch screw jaws with nuts, pins and cotters, plunger coupling with screw jaw, nut, pins and cotters, crank coupling with pins and cotters, right and left-hand screw coupling with nuts, pins and cotters, 12 inch x 12 inch wrought crank and stand with pin, cotter, bolts and washers, stand with crocodile jaw, T crank, pins, roller, disc, cotters, bolts and washers, plunger stand with round steel plunger, pin, cotters, bolts and washers, stretcher bar, Front Rod (Style B) with lugs, pins, cotters and bolts, iron plate with one end strip rivetted on and one loose, two rail braces, facing switch detector stand with stretcher bar, switch detector, pins, cotters, bolts and washers.
OA244	**Switch and Lock Movement**, complete, with two 1¼ inch wrought jaws with pins and cotters, plunger coupling with screw jaw, nut, pins and cotters, crank coupling with pins and cotters, right and left-hand screw coupling with nuts, pins and cotters, 12 inch x 12 inch wrought crank and stand with pin, cotter, bolts and washers, stand with crocodile jaw, T crank, pins, roller, disc, cotters, bolts and washers, plunger stand with round steel plunger, pin, cotters, bolts and washers, stretcher bar, Front Rod (Style B) with lugs, pins, cotters and bolts, iron plate with one end strip rivetted on and one loose, two rail braces.

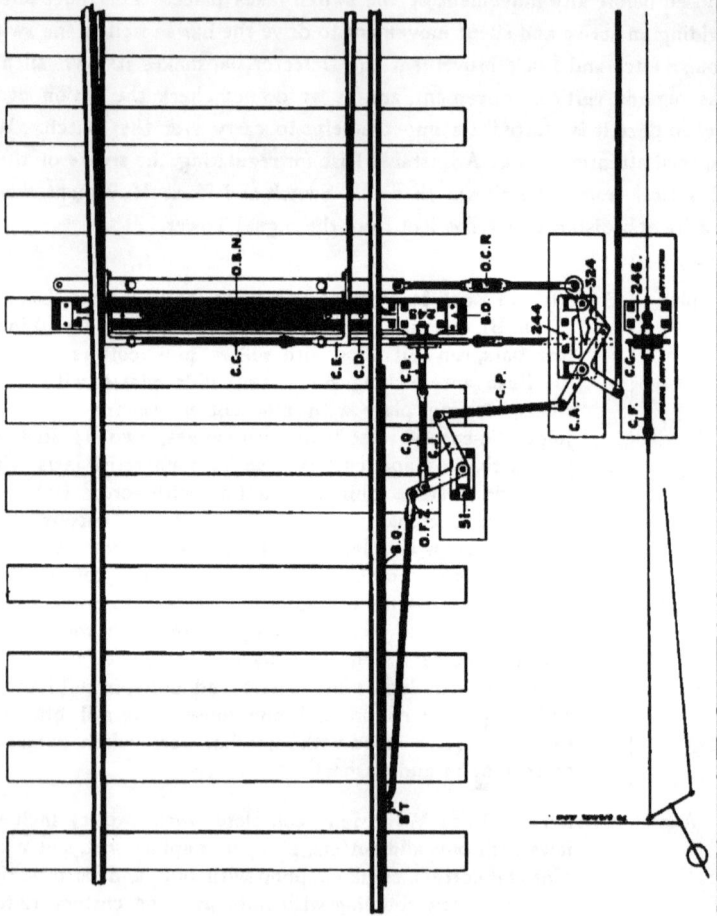

THE JOHNSON SWITCH AND LOCK MOVEMENT
O244

FACING POINT LOCKS—Continued

FITTINGS FOR SWITCH AND LOCK MOVEMENT

Order No.	
OB244	**Switch and Lock Movement Stand**, complete, with crocodile jaw, T crank, pins, roller, disc, cotters, washers and bolts.
244	**Switch and Lock Movement Stand** only.
324	**Crocodile Jaw and Bolt** only.
OCA	**T Crank**, complete, with roller and pin.
BZ	**Adjustable Disc** only.
0245	**Plunger Stand**, complete, with round steel plunger, pin, cotters, bolts and washers.
245	**Plunger Stand** only.
CB	**Round Steel Plunger** only.
OA51	**Crank Stand**, complete, with 12 inch x 12 inch wrought crank, pin, cotter, bolts and washers.
51	**Crank Stand** only.
CJ	**12 inch x 12 inch Wrought Crank** only.
0246	**Facing Switch Detector Stand**, complete, with stretcher bar (when ordering state 1 or 2-way), switch detector, pins, cotters, bolts and washers.
246	**Facing Switch Detector Stand** only.
CS	**1-way Stretcher Bar** only.
CH	**2-way Stretcher Bar** only.
CF	**Switch Detector** only.
0243	**Detector Bar (Style B)**, complete with two 18 feet bars with 5 feet 6 inch coupling bar, ten rail clips, with rollers, pins, cotters and ten pairs of Bush interlocking bolts, ten slide plates with studs and rivets, driving plate with pin, cotter, washer and rivets, eye joint, tie-bar 38 feet long with spikes.
243	**Rail Clip** only, when ordering send section of Rail.
BX	**Bush Interlocking Bolt** only.
242	**Slide Plate** only.
BQ	**Detector Bar** only.
BR	**Coupling** only for Detector Bar.

FITTINGS FOR SWITCH AND LOCK MOVEMENT

O245

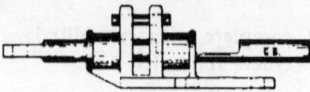

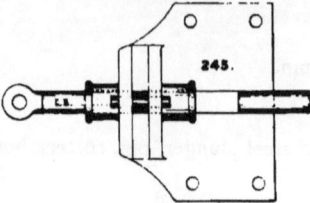

OA63

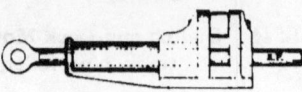

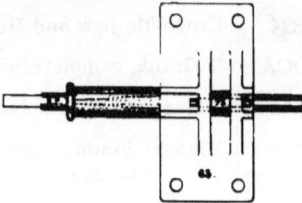

OCA

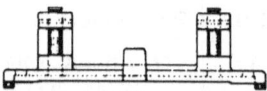

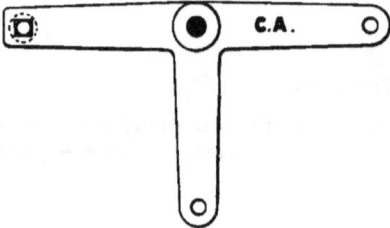

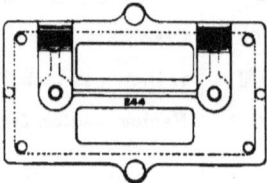

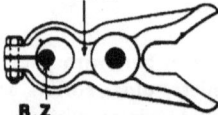

324.

B.Z

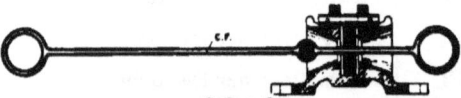

O246

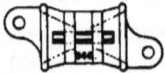

C.D.

FACING POINT LOCKS—Continued

Order No.	
BS	**Driving Plate** only for Detector Bar.
BT	**Eye Joint** only for Detector Bar.
BU	**Tie Bar** only.
OCC	**Front Rod (Style B)**, complete with two lugs, pins, cotters and bolts.
CC	**Front Rod** only.
CE	**Lugs** only.
CD	**Stretcher Bar** only.
OAY	1¼ inch **Wrought Jaw**, complete with pin and cotter.
OBN	**Iron Plate**, complete with one end strip rivetted on, and one loose.
CP	**Crank Coupling**, with pins and cotters.
CQ	**Plunger Coupling**, with screw jaw, nut, pins and cotters.
OCR	**Right and Left Hand Screw Coupling**, complete with nuts, pins and cotters.
OFZ	1¼ inch **Screw Jaw**, complete with nut, pin and cotter.
317	**Rail Braces** only.

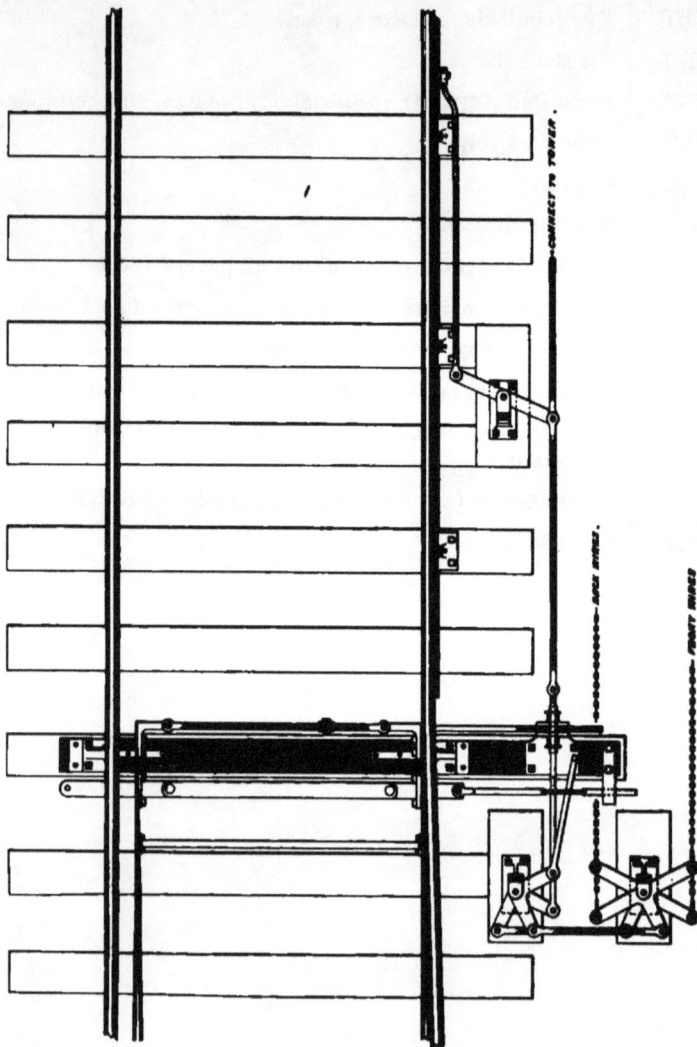

COMBINED SWITCH LOCK AND SIGNAL MOVEMENT

Pat. No. 343320, June 8, 1886

PATENT COMBINED SWITCH LOCK AND SIGNAL MOVEMENTS

The illustration shows our lock and signal movement by which one lever operates switch lock, detector bar, and two junction signals, ordinarily taking three levers to do the same work. The advantages of this arrangement are economy of levers and space in tower and certain detection of breakages in, or disruption of the connections.

The loss even of a pin will prevent a signal being lowered to the safety position until the defect has been remedied.

It will be seen that this is effected by the detector bar and locking plunger being made part of the connection which operates the signals.

We so construct the connections to the detector bar, that the lever can be reversed and signals thrown back to the danger position whilst a train stands upon the bar, the switches remaining locked until the train moves off from the detector bar. Full details of this arrangement will be furnished on application.

THE JOHNSON PATENT IMPROVED WIRE COMPENSATOR

Pat. No. 259865, 1882

O163

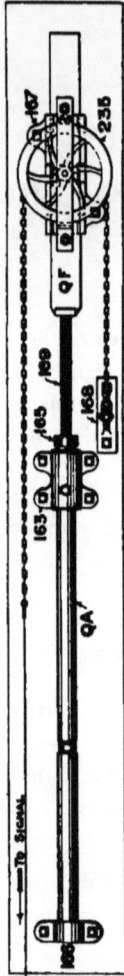

FITTINGS FOR WIRE COMPENSATOR

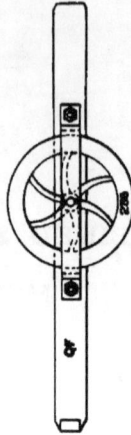

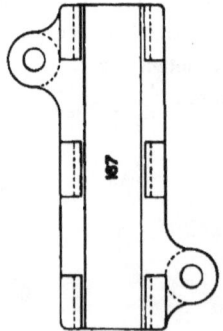

THE JOHNSON PATENT IMPROVED WIRE COMPENSATOR

This Compensator received the highest award for Compensators (The Silver Medal), at the International Inventions Exhibition, London, England, 1885.

The difficulty in working railway signals caused by the contraction and expansion of signal wires, arising from variation in temperature, is well known, and although the desirability of compensating automatically for the constantly varying length of wire has long been admitted, it may be safely asserted that hitherto, a really efficient Compensator is a want that has not been supplied. Many attempts have been made to meet this want by means of weighted appliances, but for obvious reasons these have failed. It is evident that changes of weather must be taken into consideration, since wind and dryness, as compared with calm and moisture increase the friction in all the guiding and carrying parts of the connections and in consequence of this variable friction, weights fail to act uniformly under similar state of temperature, and do not maintain constantly one length between the operating lever and the signal. Not only do such appliances fail to compensate, but they unnecessarily strain the connections, and thereby materially hasten renewal. To introduce such compensators into old connections would in many cases necessitate renewal at the same time, when the connections might continue in use for a much longer period if a compensator which adds no strain were introduced.

Our Compensator is designed to overcome these objections, and fully meets the acknowledged requirements. Its action is certain, and true to the variation of temperature, and it takes up the slack as certainly in a dry as in a moist atmosphere, it adds no strain on the connections, but rather reduces it, by maintaining a minimum of tension, not maintained by the adjusting screw, which, not being automatic, is only altered occasionally, and not exactly as the temperature varies. It is simple in construction and very durable. If by any possibility through defect or decay it should break down, no danger ensues, as under such circumstances the signal at once goes to danger. It can be inserted at trifling cost, and requires no alteration in existing arrangements. Owing to improved manufacture we have overcome all difficulties with the packing.

FITTINGS FOR WIRE COMPENSATOR

WIRE COMPENSATOR—Continued

Order No.	
0163	**Wire Compensator,** complete with stand and stuffing box, brass bush, gland nut, brass plunger, steel plug and lag screws, 1¼ inch wrought pipe with steel union, extra length of 1¼ inch pipe and sensitive rod (when ordering state distance of signal from tower), end carrier stand and lag screws, slide stand with slide brace, 10 inch wheel, pin, cotters, bolts and lag screws, 4 feet of chain, shackles, wire eyes, pins and cotter, cast lug and lag screws.

FITTINGS FOR WIRE COMPENSATOR

163	**Stand and Stuffing Box** only.
164	**Brass Bush** only.
165	**Gland Nut** only.
166	**End Carrier Stand** only.
167	**Slide Stand** only.
168	**Lug** only for fixing end of chain.
169	**Brass Plunger** only.
235	10 inch **Wheel** only.
QA	1¼ inch **Wrought Pipe** only.
QB	¾ inch **Steel Plug** only.
QC	1¼ inch **Extension Pipe** only, when ordering give length of pipe.
QD	¾ inch **Steel Union** only.
QE	**Sensitive Rod** only.
QF	**Slide Bar** only.
QG	**Slide Bar Brace** only.
CT	¾ inch **Shackle** only.
CY	**Chain** only.
DB	**Wire Eye,** oval only.

WIRE PULLEYS

0318

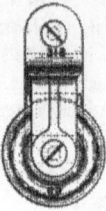 

0325

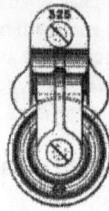 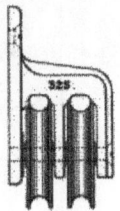

0067

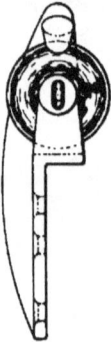 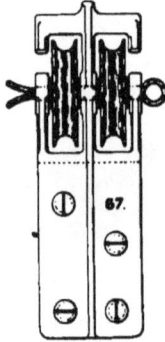

0068

 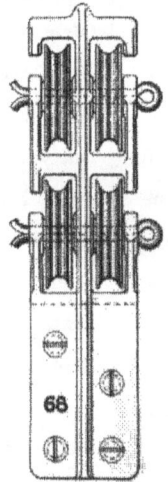

0320

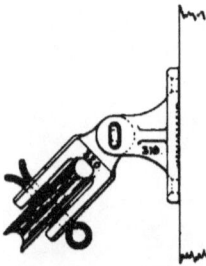 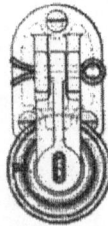

0321

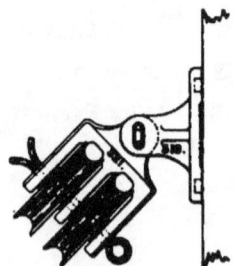

WIRE CONNECTIONS

WIRE

Order No.	
408	Seven Strand Galvanized Wire.
409	Single Strand Steel Wire.
	When ordering state number of feet required.

WIRE PULLEYS

067	Style A, 1-way Wire Pulley, complete, with stand, pulley, cotter and screws.
0067	Style A, 2-way Wire Pulley, complete, with stand, two pulleys, cotter and screws.
068	Style A, 3-way Wire Pulley, complete, with stand, three pulleys, cotters and screws.
0068	Style A, 4-way Wire Pulley, complete, with stand, four pulleys, cotters and screws.
67	Style A, 1 and 2-way Wire Pulley Stand only.
68	Style A, 3 and 4-way Wire Pulley Stand only.
0318	Style A, 1-way Side Wire Pulley, complete, with stand, pulley and screws.
0325	Style A, 2-way Side Wire Pulley, complete, with stand, two pulleys and screws.
318	Style A, 1-way Side Wire Pulley Stand only.
325	Style A, 2-way Side Wire Pulley Stand only.
0320	Style A, 1-way Angle Wire Pulley, complete with stand, bracket, pulley, cotters and screws.
0321	Style A, 2-way Angle Wire Pulley, complete, with stand, bracket, two pulleys, cotters and screws.
319	Style A, Stand for 1 and 2-way Angle Wire Pulleys only.
320	Style A, Bracket for 1-way Angle Wire Pulley only.
321	Style A, Bracket for 2-way Angle Wire Pulley only.
69	Wire Pulley only.

WHEELS

0226

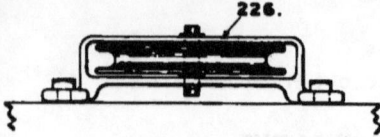

0227

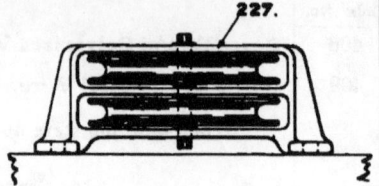

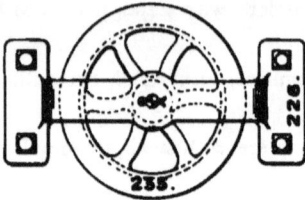

0228

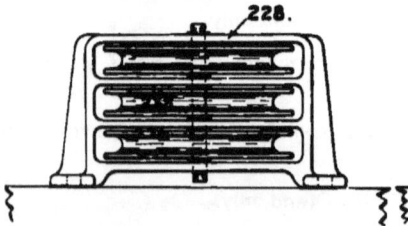

0229

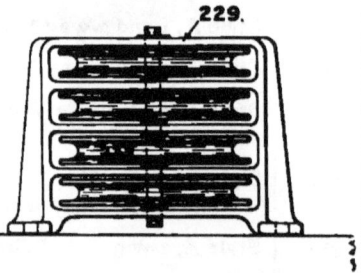

0138

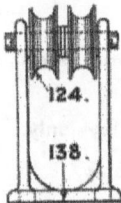

0140

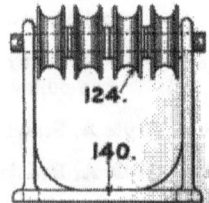

WIRE CONNECTIONS—Continued

WHEELS

For Vertical Wheels see pages 37, 38 and 39

Order No.	
0214	1-way 6 inch Hor. Wheel Stand, complete, with wheel, pin, cotters and lag screws.
0215	2-way 6 inch Hor. Wheel Stand, complete, with two wheels, pin, cotters and lag screws.
0216	3-way 6 inch Hor. Wheel Stand, complete, with three wheels, pin, cotters and lag screws.
0217	4-way 6 inch Hor. Wheel Stand, complete, with four wheels, pin, cotters and lag screws.
214	1-way 6 inch Hor. Wheel Stand only.
215	2-way 6 inch Hor. Wheel Stand only.
216	3-way 6 inch Hor. Wheel Stand only.
217	4-way 6 inch Hor. Wheel Stand only.
234	6 inch Wheel only.
0226	1-way 10 inch Hor. Wheel Stand, complete, with wheel, pin, cotters and lag screws.
0227	2-way 10 inch Hor. Wheel Stand, complete, with two wheels, pin, cotters and lag screws.
0228	3-way 10 inch Hor. Wheel Stand, complete, with three wheels, pin, cotters and lag screws.
0229	4-way 10 inch Hor. Wheel Stand, complete, with four wheels, pin, cotters and lag screws.
226	1-way 10 inch Hor. Wheel Stand only.
227	2-way 10 inch Hor. Wheel Stand only.
228	3-way 10 inch Hor. Wheel Stand only.
229	4-way 10 inch Hor. Wheel Stand only
235	10 inch Wheel only.

ONE ARM BLOCK STATION SIGNAL
· FIXED OPPOSITE TOWER

Order No.	
0171	**One Arm Block Station Signal**, fixed opposite Tower, complete, with pine post 26 feet long, wrought-iron ladder with round rungs, four stays, bolts and lag screws, semaphore bearing with blade grip, two rings, back light spectacle, two glasses, bolts and lag screws, semaphore spindle with nuts and cotters, ash blade with bolts, semaphore eye joint, wire eyes, 6 feet seven strand wire, split links, and 4 feet of chain, 6 inch side wheel, with stand, pin, cotters, and lag screws, standard lamp with bracket and lag screws.

FITTINGS

GJ	**Post** only, pine 26 feet long.
DN	**Wrought Ladder**, with round rungs and four stays.
DO	**Ash Blade** only.
DQ	**Semaphore Spindle** only.
DR	**Semaphore Eye Joint** only.
DB	**Oval Wire Eye** only.
CZ	**Ordinary Split Link** only.
CY	**Chain** only (when ordering state number of feet).
408	**Wire** only, seven strand galvanized.
410	**Standard Lamp** only.
DM	**Lamp Bracket** only.
197	**Semaphore Bearing** only.
199	**Blade Grip** only.
101	**Ring** only, for Blade Grip.
TA	**7 inch Ruby Glass** only, for Blade Grip.
102	**Back Light Spectacle** only.
103	**Ring** only for Back Spectacle.
TB	**Blue Glass** only, for Back Spectacle.
214	**1-way 6 inch Side Wheel Stand** only.
234	**6 inch Wheel** only.

TWO ARM BLOCK STATION SIGNAL
FIXED OPPOSITE TOWER
Illustrated on Page 80

Order No.	
0200	**Two Arm Block Station Signal,** fixed opposite tower, complete, with pine post 26 feet long, wrought-iron ladder with round rungs, four stays, bolts and lag screws, two semaphore bearings with right and left hand blade grips, two rings, two ruby glasses, bolts and lag screws, semaphore spindle with nuts and cotters, two ash blades with bolts, two semaphore eye joints, wire eyes, 12 feet seven strand wire, split links, and 8 feet of chain, two 6 inch side wheels with two stands, pins, cotters and lag screws, standard lamp with bracket and lag screws.

FITTINGS

GJ	**Post** only, pine 26 feet long.
DN	**Wrought Ladder,** with round rungs and four stays.
DO	**Ash Blade** only.
DQ	**Semaphore Spindle** only.
DR	**Semaphore Eye Joint** only.
DB	**Oval Wire Eye** only.
CZ	**Ordinary Split Link** only.
CY	**Chain** only (when ordering state number of feet).
408	**Wire** only, seven strand galvanized.
410	**Standard Lamp** only.
DM	**Lamp Bracket** only.
197	**Semaphore Bearing** only.
198	**Right Hand Blade Grip** only.
199	**Left Hand Blade Grip** only.
101	**Ring** only for Blade Grip.
TA	7 inch **Ruby Glass** only.
214	1-way 6 inch **Side Wheel Stand** only.
234	6 inch **Wheel** only.

TWO ARM BLOCK STATION SIGNAL
FIXED OPPOSITE TOWER
0200

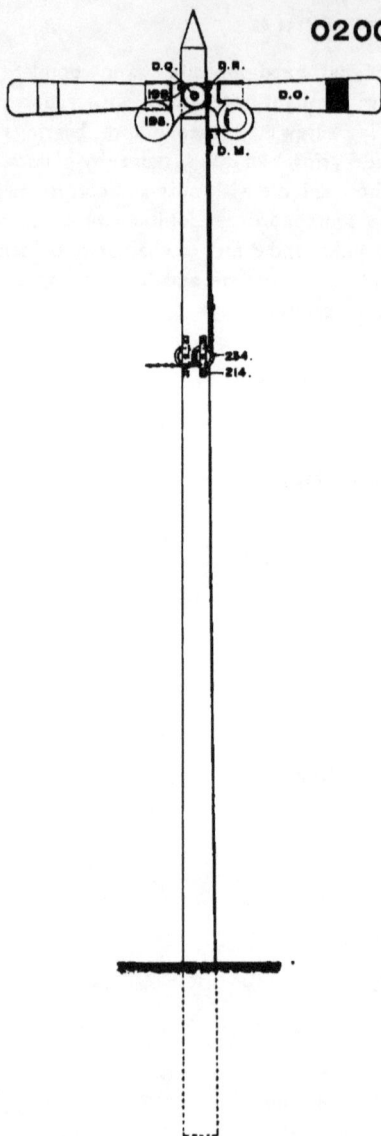
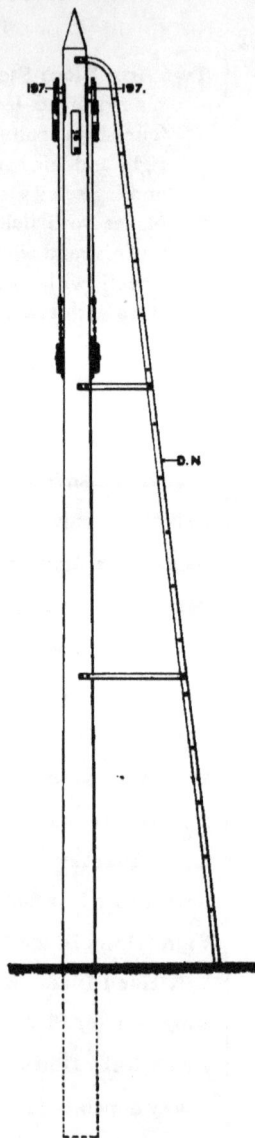

ONE ARM BLOCK STATION SIGNAL
FOR
DISTANT WORKING
0196

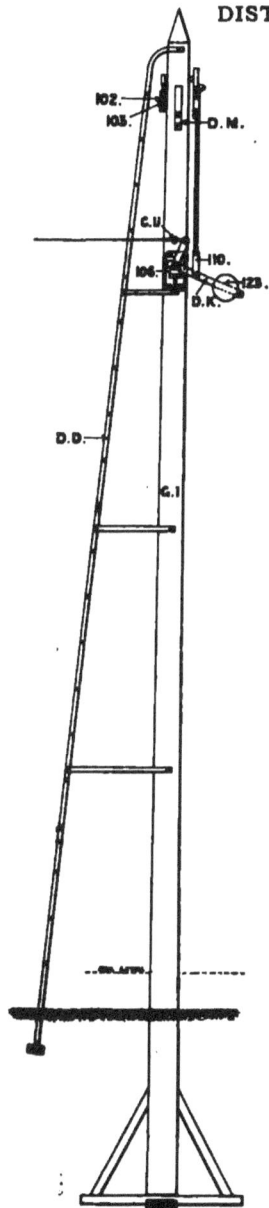

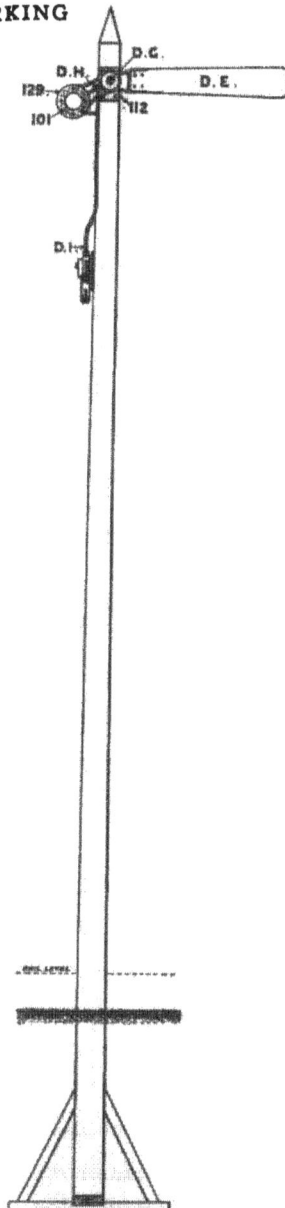

ONE ARM BLOCK STATION SIGNAL

FOR DISTANT WORKING

Illustrated on Page 81

Order No.	
0196	**One Arm Block Station Signal, for Distant Working,** complete, with pine post 28 feet long, with cross bottom pieces, braces, two hook bolts, washers and lag screws, wrought-iron ladder with round rungs, six stays, bolts and lag screws, semaphore bearing with blade grip, back light spectacle, two rings, two glasses, bolts and lag screws, semaphore spindle with nuts and cotters, ash blade with bolts (for home or distant signals, when ordering, state which is required), semaphore eye joint with four feet of $\frac{3}{4}$ inch pipe for down rod, screw end and nut, $\frac{3}{4}$ inch screw jaw, pin and cotters, balance lever with weight, stand, pins, cotters, shackle, bolt and lag screws, standard lamp with bracket and lag screws.

FITTINGS

GI	**Post** (pine), 28 feet long with two cross bottom pieces and four braces.
DD	**Wrought Ladder,** with round rungs and six stays.
DE	**Ash Blade** only, for Home Signal.
DF	**Ash Blade** only, for Distant Signal.
DG	**Semaphore Spindle** only.
112	**Semaphore Bearing** only.
129	**Blade Grip** only.
101	**Ring** only, for blade grip.
TA	7 inch **Ruby Glass** only for blade grip.
102	**Back Light Spectacle** only.
103	**Ring** only, for back light spectacle.
TB	$3\frac{1}{2}$ inch **Blue Glass** only, for back light spectacle.
DM	**Lamp Bracket** only.

ONE ARM BLOCK STATION SIGNAL, FOR DISTANT WORK-
ING—Continued

Order No.	
410	**Standard Lamp** only.
DH	**Semaphore Eye Joint** only.
GG	¾ inch **Pipe** only, for down rod.
DI	**Screw End and Nut** only, for ¾ inch screw jaw.
110	¾ inch **Screw Jaw** only.
106	**Balance Lever Stand** only.
DK	**Balance Lever** only.
123	**Balance Weight** only
CU	⅝ inch **Shackles** only.
DW	**Hook Bolts** only.

ONE ARM C PATTERN SIGNAL

PIPE CONNECTED

Illustrated on Page 84

0313	**One Arm C Pattern Signal, Pipe connected,** complete, with pine post 28 feet long, with cross bottom pieces, braces, two hook bolts, washers and lag screws, wrought-iron ladder with round rungs, six stays, bolts and lag screws, semaphore bearing, with blade grip, back light spectacle, two rings, two glasses, bolts and lag screws, semaphore spindle with nuts and cotters, ash blade with bolts (for Home or Distant Signals, when ordering state which is required), semaphore eye joint with ¾ inch pipe for down rod, sleeve, three 1-way pipe guides, screw end and nut, ¾ inch screw jaw, bolts, pins, cotters and lag screws, balance lever with weight, stand, pin, cotter, bolt and lag screws, standard lamp with bracket and lag screws.

FITTINGS

GK	**Post** (pine), 28 feet long with two cross bottom pieces and four braces.
DS	**Wrought Ladder,** with round rungs and six stays.
DT	**Ash Blade** only, for Home Signal.

ONE ARM C PATTERN SIGNAL

PIPE CONNECTED

O313

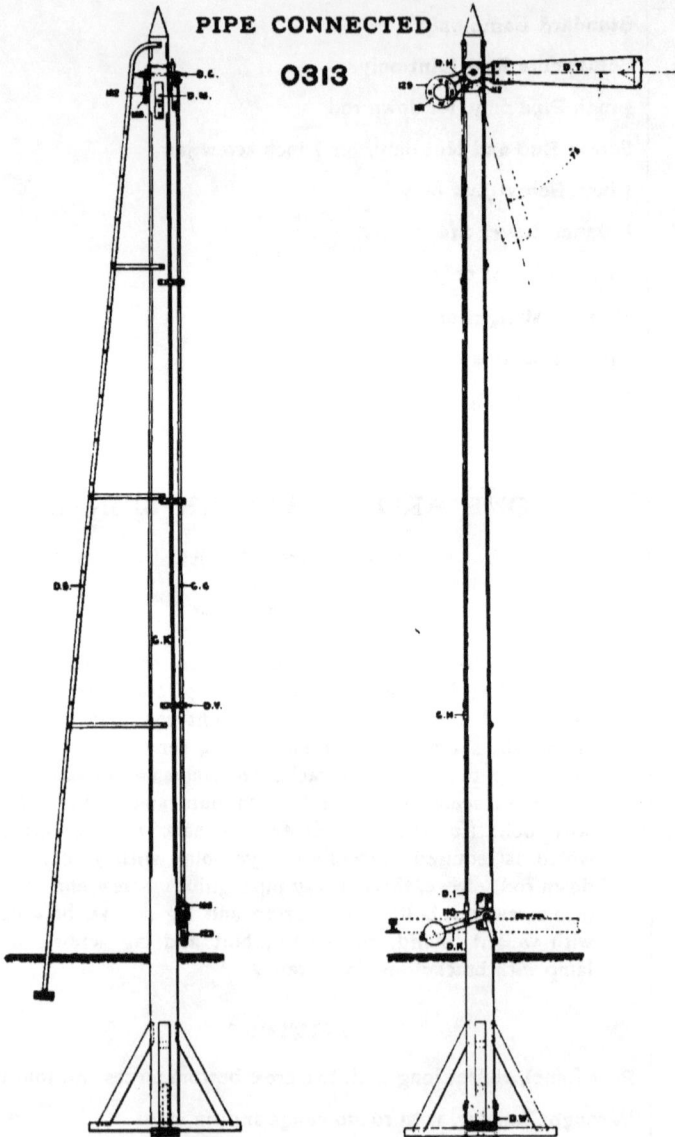

ONE ARM C PATTERN SIGNAL, PIPE CONNECTED—Continued

Order No.	
DU	Ash Blade only, for Distant Signal.
SG	Semaphore Spindle only.
112	Semaphore Bearing only.
129	Blade Grip only.
101	Ring only, for blade grip.
TA	7 inch Ruby Glass only, for blade grip.
102	Back Light Spectacle only.
103	Ring only, for back light spectacle.
TB	3½ inch Blue Glass only.
DM	Lamp Bracket only.
410	Standard Lamp only.
DH	Semaphore Eye Joint only.
GG	¾ inch Pipe only, for down rod (when ordering state number of feet required).
DI	Screw End and Nut only, for ¾ inch screw jaw.
110	¾ inch screw Jaw only.
GH	¾ inch Pipe Sleeve only.
DV	1-way Pipe Guides only.
106	Balance Lever Stand only.
DK	Balance Lever only.
123	Balance Weight only.
DW	Hook Bolts only.

ONE ARM C PATTERN SIGNAL

WIRE CONNECTED, WITH SIGNAL MOVEMENT (STYLE A)

Order No.

0416	One Arm C, Pattern Signal, Wire connected with Signal Movement (Style A), complete, with pine post 28 feet long, with cross bottom pieces, braces, two hook bolts, washers and lag screws, wrought-iron ladders with round rungs, six stays, bolts, and lag screws, semaphore bearing with blade grip, back light spectacle, two rings, two glasses, bolts and lag screws, semaphore spindle with nut and cotters, ash blade with bolts, (for home or distant signals, when ordering state which is required), semaphore eye joint, with 2 feet 8 inches of $\frac{1}{2}$ inch pipe for down rod, screw end and nut, $\frac{1}{2}$ inch screw jaw, pins and cotters, signal movement stand (Style A), with wheel, crank, roller lever, balance weight, pins, cotters, bolts and lag screws, hanger with five drop weights, 13 feet of chain, 26 feet of wire, one $\frac{1}{2}$ inch shackle, eight wire eyes, ring, six split links, pins and cotters, 2-way signal wheel stand with 6 inch and 10 inch wheels, pins, cotters and lag screws, standard lamp with bracket and lag screws.

FITTINGS

GK	Post (pine), 28 feet long with two cross bottom pieces and four braces.
DS	Wrought Ladder, with round rungs and six stays.
DT	Ash Blade only, for Home Signal.
DU	Ash Blade only, for Distant Signal.
DG	Semaphore Spindle only.
112	Semaphore Bearing only.
129	Blade Grip only.
101	Ring only, for blade grip.
TA	7 inch Ruby Glass only, for blade grip.
102	Back Light Spectacle only.
103	Ring only, for back light spectacle.
TB	$3\frac{1}{2}$ inch Blue Glass only, for back light spectacle.
DM	Lamp Bracket only.

ONE ARM C PATTERN SIGNAL, WIRE CONNECTED WITH SIGNAL MOVEMENT (STYLE A)—Continued

Order No.	
410	Standard Lamp only.
DH	Semaphore Eye Joint only.
GG	¼ inch Pipe only, for down rod.
DI	Screw End and Nut only for ¼ inch screw jaw.
110	¾ inch Screw Jaw only.
22	Signal Movement Stand (Style A) only.
23	Wheel only, for signal movement.
24	Crank only, for signal movement.
FA	Roller only, for signal movement.
EL	Lever only, for signal movement
123	Balance Weight only for signal movement.
EM	Hanger only.
130	Drop Weights only, 6lbs
131	Drop Weights only, 12lbs.
CT	½ inch Shackles only.
DB	Oval Wire Eyes only.
CZ	Ordinary Split Links only.
408	Wire, seven strand galvanized only.
CY	Chain only.
RK	2 inch Wrought Ring only.
230	2-way Signal Wheel Stand only.
234	6 inch Wheel only for 2-way signal wheel.
235	10 inch Wheel only for 2-way signal wheel.
DW	Hook Bolts only.

ONE ARM C PATTERN SIGNAL

WIRE CONNECTED WITH SIGNAL MOVEMENT (STYLE B)

Order No.	
0417	One Arm C Pattern Signal, Wire connected with Signal Movement (Style B), complete, with pine post 28 feet long, with cross bottom pieces, braces, two hook bolts, washers and lag screws, wrought-iron ladder with round rungs, six stays, bolts and lag screws, semaphore bearing with blade grip, back light spectacle, two rings, two glasses, bolts and lag screws, semaphore spindle with nuts and cotters, ash blade with bolts (for home or distant signals, when ordering state which is required), semaphore eye joint with 2 feet 8 inches of $\frac{3}{4}$ inch pipe for down rod, screw end and nut, $\frac{3}{4}$ inch screw jaw, pins and cotters, signal movement stand (Style B), with crank, three armed lever, roller, balance weight, pins, cotters and lag screws, 30 feet of wire, with two $\frac{3}{8}$ inch shackles, four wire eyes, two split links, 5 feet of chain, pins and cotters, 2-way signal wheel stand with 6 inch and 10 inch wheels, pins, cotters and lag screws, standard lamp with bracket and lag screws.

FITTINGS

GK	Post (pine), 28 feet long with two cross bottom pieces and four braces.
DS	Wrought Ladder, with round rungs and six stays.
DT	Ash Blade only, for Home Signal.
DU	Ash Blade only, for Distant Signal.
DG	Semaphore Spindle only.
112	Semaphore Bearing only.
129	Blade Grip only.
101	Ring only, for blade grip.
TA	7 inch Ruby Glass only, for blade grip.
102	Back Light Spectacle only.
103	Ring only, for back light spectacle.
TB	$3\frac{1}{2}$ inch Blue Glass only, for back light spectacle.
DM	Lamp Bracket only.

ONE ARM C PATTERN SIGNAL, WIRE CONNECTED WITH SIGNAL MOVEMENT (STYLE B)—Continued

Order No.	
410	Standard Lamp only.
DH	Semaphore Eye Joint only.
GG	¼ inch Pipe only, for down rod.
DI	Screw End and Nut only, for ¾ inch screw jaw.
110	¾ inch Screw Jaw only.
184	Signal Movement Stand (Style B) only.
183	Crank only, for signal movement.
EN	Three Armed Balance Lever only, for signal movement.
FA	Roller only, for balance lever—for signal movement.
123	Balance Weight only, for signal movement.
408	Wire seven strand galvanized only.
CT	¼ inch Shackles only.
DB	Oval Wire Eyes only.
CZ	Ordinary Split Links only.
CY	Chain only.
230	2-way Signal Wheel Stand only.
234	6 inch Wheel only, for 2-way signal wheel.
235	10 inch Wheel only, for 2-way signal wheel.
DW	Hook Bolts only.

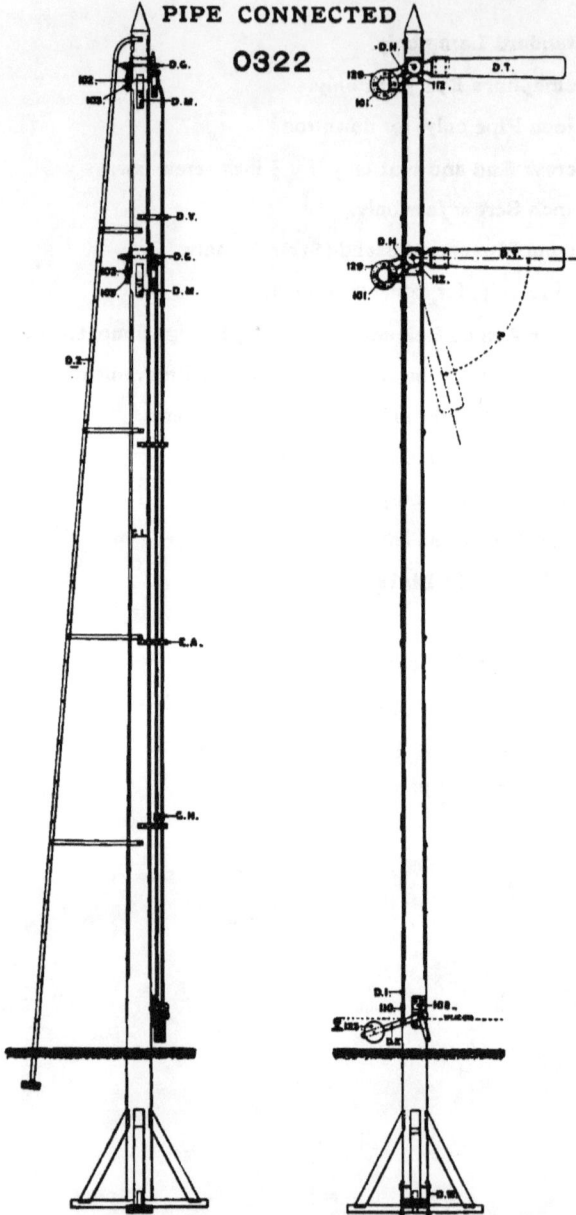

TWO ARM C PATTERN SIGNAL
PIPE CONNECTED
0322

TWO ARM C PATTERN SIGNAL

PIPE CONNECTED

Order No.	
0322	**Two Arm C Pattern Signal, Pipe connected,** complete, with pine post 34 feet long, with cross bottom pieces, braces, two hook bolts, washers and lag screws, wrought-iron ladder with round rungs, eight stays, bolts and lag screws, two semaphore bearings with two blade grips, two back light spectacles, four rings, four glasses, bolts and lag screws, two semaphore spindles with nuts and cotters, two ash blades with bolts (for home or distant signals, when ordering state which is required), two semaphore eye joints with $\frac{1}{4}$ inch pipe, for down rods (when ordering give length of pipe required), three sleeves, one 1-way and three 2-way pipe guides, two screw ends and nuts, two $\frac{1}{4}$ inch screw jaws, bolts, pins, cotters and lag screws, 2-way balance lever stand with two balance levers, two weights, pin, cotter, bolts and lag screws, two standard lamps with two brackets and lag screws.

FITTINGS

GL	Post, pine 34 feet long with two cross bottom pieces and four braces.
DZ	Wrought Ladder, with round rungs and eight stays.
DT	Ash Blade only, for Home Signal.
DU	Ash Blade only, for Distant Signal.
DG	Semaphore Spindle only.
112	Semaphore Bearing only.
129	Blade Grip only.
101	Ring only, for blade grip.
TA	7 inch Ruby Glass only, for blade grip.
102	Back Light Spectacle only.
103	Ring only, for back light spectacle.
TB	3½ inch Blue Glass only.
DM	Lamp Bracket only.

TWO ARM C PATTERN SIGNAL—Continued

Order No.	
410	Standard Lamp only.
DH	Semaphore Eye Joint only.
GG	¾ inch Pipe only, for down rods (when ordering state number of feet required).
DV	1-way Pipe Guide only.
EA	2-way Pipe Guide only.
GH	¾ inch Pipe Sleeve only.
DI	Screw End and Nut only, for ¾ inch screw jaw.
110	¾ inch Screw Jaw only.
108	2-way Balance Lever Stand only.
DK	Balance Lever only.
123	Balance Weight only.
DW	Hook Bolts only.
0418	**TWO ARM C PATTERN SIGNAL, Wire Connected with Signal Movement (Style A).**
0419	**TWO ARM C PATTERN SIGNAL, Wire Connected with Signal Movement (Style B).**

TWO ARM C PATTERN BRACKET SIGNAL

PIPE CONNECTED

Illustrated on Page 94

Order No.	
0316	Two Arm C Pattern Bracket Signal, Pipe connected, complete, with pine posts, one 25 feet and two 10 feet long, with top braces, oak joists, flooring, cross bottom pieces, bottom braces, two hook bolts, bolts, washers and lag screws, wrought-iron ladder with round rungs, four braces, bolts, washers and lag screws, front, back and side handrails with stays, nuts, washers and lag screws, two semaphore bearings with two blade grips, two back light spectacles, four rings, four glasses, bolts and lag screws, two semaphore spindles with nuts and cotters, two ash blades with bolts (for home or distant signals, when ordering state which is required) two semaphore eye joints with $\frac{1}{4}$ inch pipe for down rods (when ordering give length of pipe required), two 1-way pipe guides, bolts and lag screws, eight $\frac{1}{4}$ inch wrought jaws, two screw ends and nuts, two $\frac{1}{4}$ inch screw jaws, pins and cotters, three pin plates with four 6 inch cranks, pins, cotters, washers and lag screws, 2-way balance lever stand, with two balance levers, two weights, pin, cotter, bolts and lag screws, two Standard lamps with two brackets and lag screws.

FITTINGS

GN	Posts (pine) one 25 feet and two 10 feet long with four top braces, two oak joists, flooring, four bottom braces and two cross bottom pieces.
EC	Wrought Ladder with round rungs and four braces.
ED	Side Handrail only, wrought-iron.
EF	Back Handrail only.
EG	Front Handrail only.
EH	Back Handrail, Stays only.
DT	Ash Blade only, for Home Signal.
DU	Ash Blade only, for Distant Signal.

TWO ARM C PATTERN BRACKET SIGNAL
PIPE CONNECTED   O316

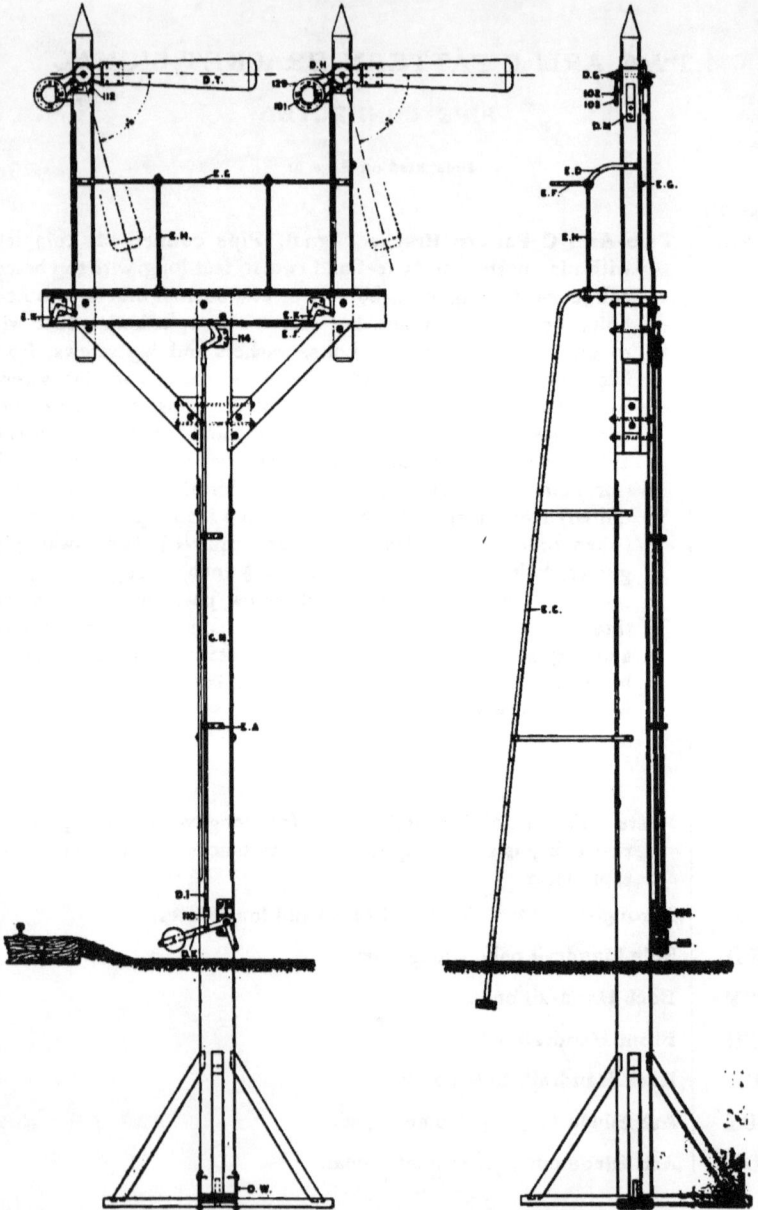

SIGNAL FITTINGS

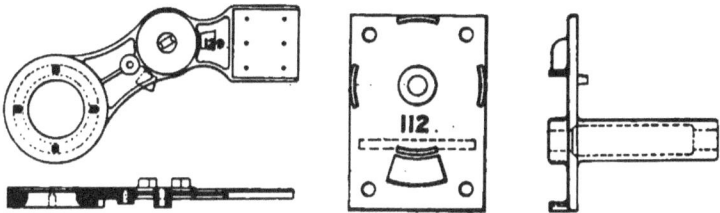

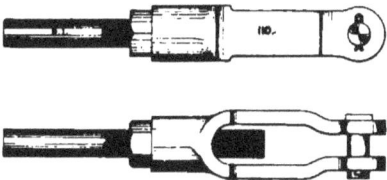

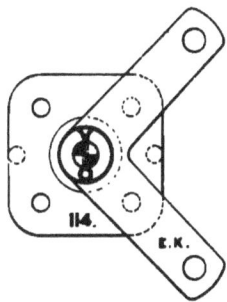

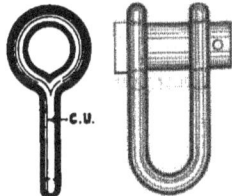

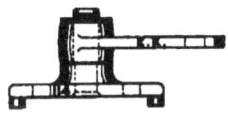

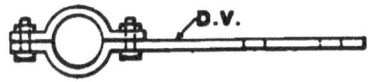

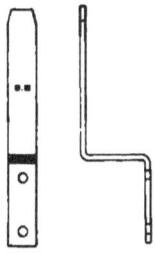

TWO ARM C PATTERN BRACKET SIGNAL—Continued

Order No.	
DG	Semaphore Spindle only.
112	Semaphore Bearing only.
129	Blade Grip only.
101	Ring only, for blade grip.
TA	7 inch Ruby Glass only, for blade grip.
102	Back Light Spectacle only.
103	Ring only, for back light spectacle.
TB	3½ inch Blue Glass only.
DM	Lamp Bracket only.
410	Standard Lamp only.
DH	Semaphore Eye Joint only.
GG	¾ inch Pipe only, for down rods (when ordering state number of feet required).
EA	2-way Pipe Guides only.
114	Pin Plates only.
EK	Wrought Cranks only, 6 inches x 6 inches.
EJ	¾ inch Wrought Jaws only.
DI	Screw End and Nut only, for ¾ inch screw jaw.
110	¾ inch Screw Jaw only.
108	2-way Balance Lever Stand only
DK	Balance Lever only.
123	Balance Weight only.
DW	Hook Bolts only.
0420	**TWO ARM C PATTERN BRACKET SIGNAL,** Wire Connected with Signal Movement (Style A).
0421	**TWO ARM C PATTERN BRACKET SIGNAL,** Wire Connected with Signal Movement (Style B).

SIGNAL MOVEMENTS

Illustrated on Page 98

We have for some time successfully used a compensating movement in Ground Disc Signals, and have lately applied the same principle to Semaphore Signals, with good results. This device which we term a Signal Movement we make in two Styles (A and B) as shown in the illustration.

Order No.	
	## STYLE **A**
022	**Signal Movement**, complete, with stand, wheel, crank, roller, balance lever, weight, hanger, five drop weights, five feet of chain, pins, bolt, cotters and lag screws.
22	**Stand** only.
23	**Wheel** only.
24	**Crank** only.
FA	**Roller** only.
EL	**Balance Lever** only.
123	**Balance Weight** only.
EM	**Hanger** only.
130	**Drop Weight** only 6 lbs.
131	**Drop Weight** only 12 lbs.
CY	**Chain** only.
	## STYLE **B**
0184	**Signal Movement**, complete with stand, crank, three arm balance lever, roller, weight, pins, cotters and lag screws.
184	**Stand** only.
183	**Crank** only.
EN	**Three Arm Balance Lever** only.
FA	**Roller** only.
123	**Balance Weight** only.

SIGNAL MOVEMENTS

STYLE B, O184

STYLE A, O22

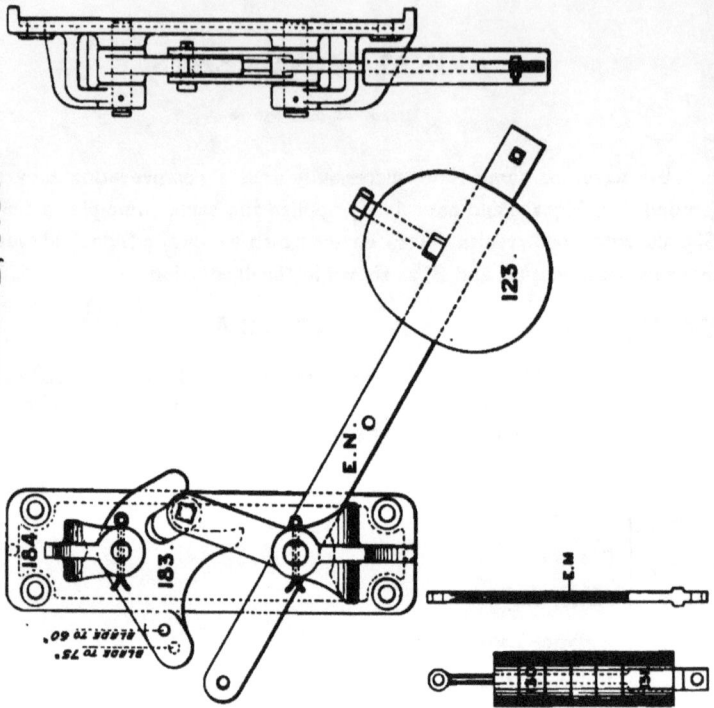

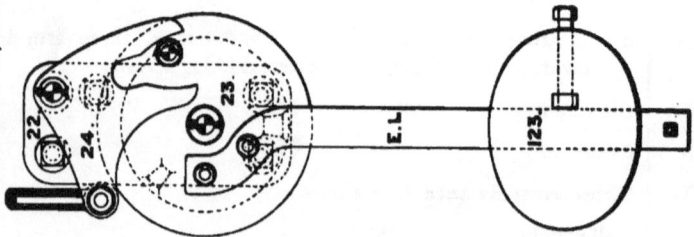

SIGNAL WHEELS

0230

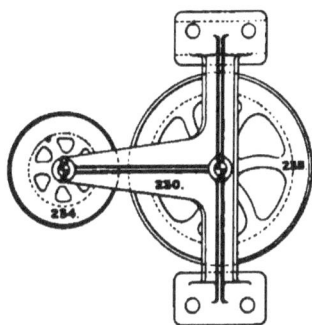 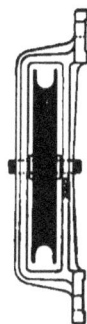

0231

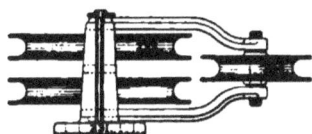 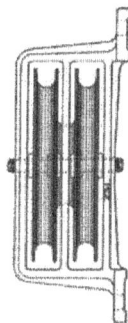

0232

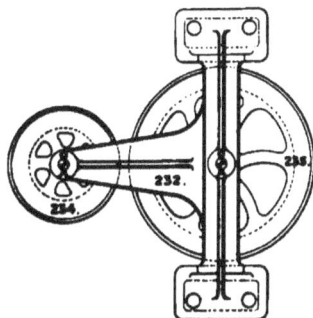 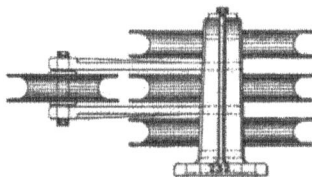

SIGNAL WHEELS

Illustrated on Page 99

Order No.	
0230	2-way Signal Wheel for One Arm Signal, complete, with stand, one 6 inch wheel and one 10 inch wheel, pins, cotters and lag screws.
0231	3-way Signal Wheel for Two Arm Signal, complete, with stand, one 6 inch wheel and two 10 inch wheels, pins, cotters and lag screws.
0232	4-way Signal Wheel for Three Arm Signal, complete, with one 6 inch wheel and three 10 inch wheels, pins, cotters and lag screws.
230	2-way Signal Wheel Stand only.
231	3-way Signal Wheel Stand only.
232	4-way Signal Wheel Stand only.
234	6 inch Wheel only.
235	10 inch Wheel only.

THE JOHNSON PATENT SIGNAL SLOT

Illustrated on Page 102

This is a device whereby two or more men may control a signal arm. This becomes necessary where the systems from two separate towers overlap, and in similar cases. It requires the concerted action of the two or more men to give the "Clear" Signal, but any one of the men may throw the Signal to Danger. The special advantage of the Johnson Slot is that the Signal Arm is forced both "On" and "Off," and the Arm is not weighted to all clear, as in other slots. At the same time when the balance levers do not work to their full stroke, the loss communicated to the blade is not the sum of such loss, but the action of that balance lever which has the least movement is communicated to the blade. Our "Slot" is Standard on important railways both in Europe and America.

Order No.	
0438	2-way Slot for One Arm Signal, complete, with stand, cover, bolts and lag screws, two rollers, swing jaw, ⅟ inch screw end and nut, two slide bars, two malleable couplings, pins and cotters, 2-way balance lever stand with two balance levers, two weights, pin, bolts and lag screws.
00438	2-way Slot for Two Arm Signal, complete, with stand, cover, bolts and lag screws, two rollers, swing jaw, ⅟ inch screw end and nut, ⅟ inch screw jaw, two slide bars, two malleable couplings, pins and cotters, 2-way balance lever stand, with two balance levers, two weights, pin, bolts and lag screws.

FITTINGS

438	Stand only.
439	Cover only.
440	Swing Jaw only.
SP	Roller only.
DI	⅟ inch Screw End and Nut only.
110	⅟ inch Screw Jaw only.
SN	Slide Bar only, for Two Arm Signals.
SO	Slide Bar only.
441	Malleable Coupling only.
108	2-way Balance Lever Stand only.
SQ	Balance Lever only.
123	Balance Weight only.

THE JOHNSON PATENT SIGNAL SLOT

Pat. No. 334232, Jan. 12 1886

00438

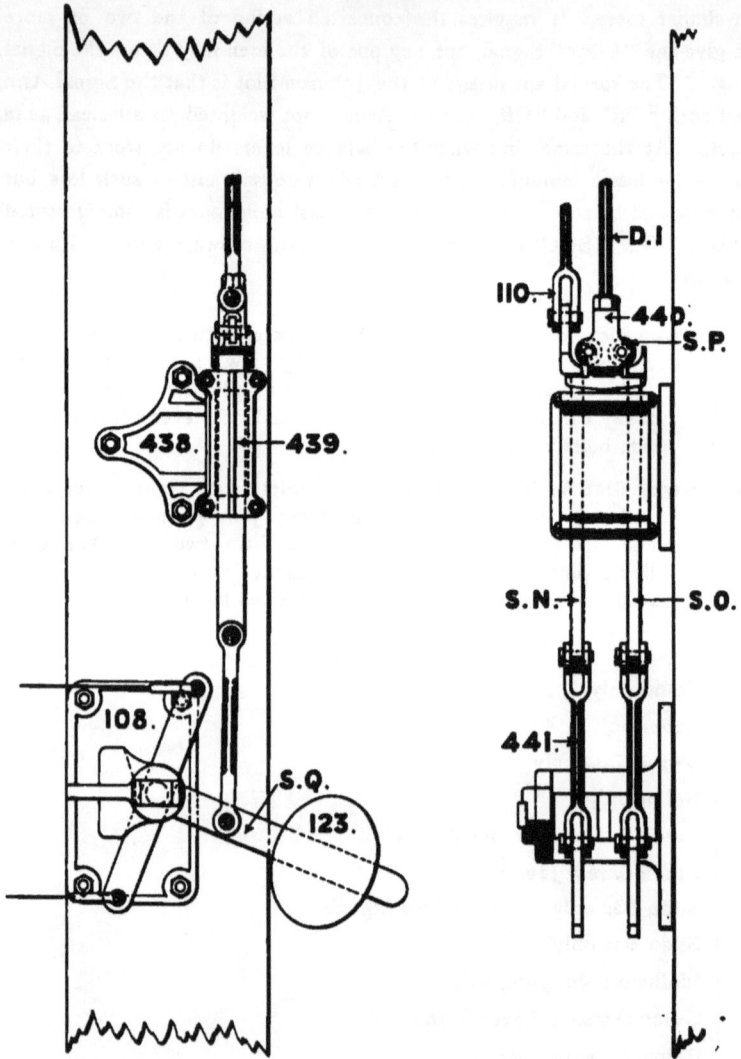

DWARF SIGNALS

Mr. Johnson invented and introduced the "Dwarf Semaphore Signal" for shifting and other slow movements. It has almost entirely superseded the Pot Signal for similar purposes, and is Standard on several railroads both in America and Europe.

Order No.	
0115	**One Arm Dwarf Signal**, complete, with stand and lag screws, 3 feet 3 inches of 2¼ inch pipe, top bearing and set screws, semaphore spindle with nuts and cotters, semaphore blade grip with back light spectacle, two rings, two glasses and bolts, ash blade with bolts, 1 foot 9 inches of ⅝ inch solid down rod with eye joint, ¾ inch screw jaw with nut, pin and cotter, balance lever with weight, pin and bolts, Standard dwarf signal lamp with bracket and set screws. Illustrated on page 104.
OA115	**Two Arm Dwarf Signal**, complete, with stand and lag screws, 4 feet 9 inches of 2¼ inch pipe, centre and top bearings and set screws, two semaphore spindles with nuts and cotters, two semaphore blade grips with two back light spectacles, four rings, four glasses and bolts, two ash blades with bolts, two ⅝ inch solid down rods with eye joints 3 feet 3 inches and 1 foot 9 inches in length, two ¾ inch screw jaws with nuts, pins and cotters, two balance levers, with two weights, pin and bolts, two Standard dwarf signal lamps with two brackets and set screws. Illustrated on page 105.
OB115	**Three Arm Dwarf Signal**, complete, with stand and lag screws, 6 feet 3 inches of 2¼ inch pipe, top bearing, two centre bearings and set screws, three semaphore spindles with nuts and cotters, three semaphore blade grips with three back light spectacles, six rings, six glasses and bolts, three ash blades with bolts, three ⅝ inch solid down rods with eye joints 4 feet 9 inches, 3 feet 3 inches, and 1 foot 9 inches in length, three ¾ inch screw jaws with nuts, pins and cotters, three balance levers with three weights, pin and bolts, three Standard dwarf signal lamps with three brackets and set screws.

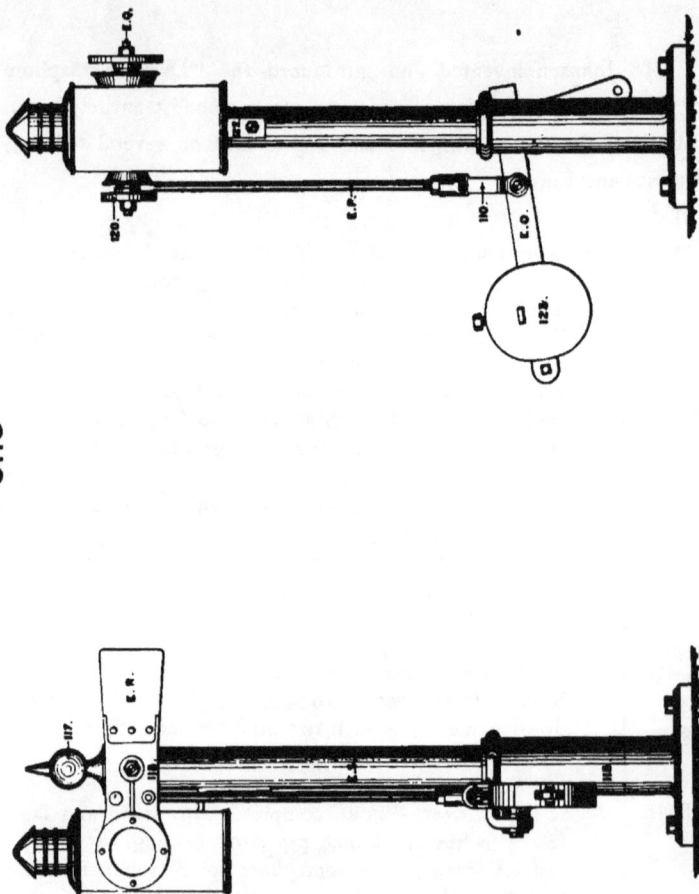

ONE ARM DWARF SIGNAL

O115

TWO ARM DWARF SIGNAL

OA115

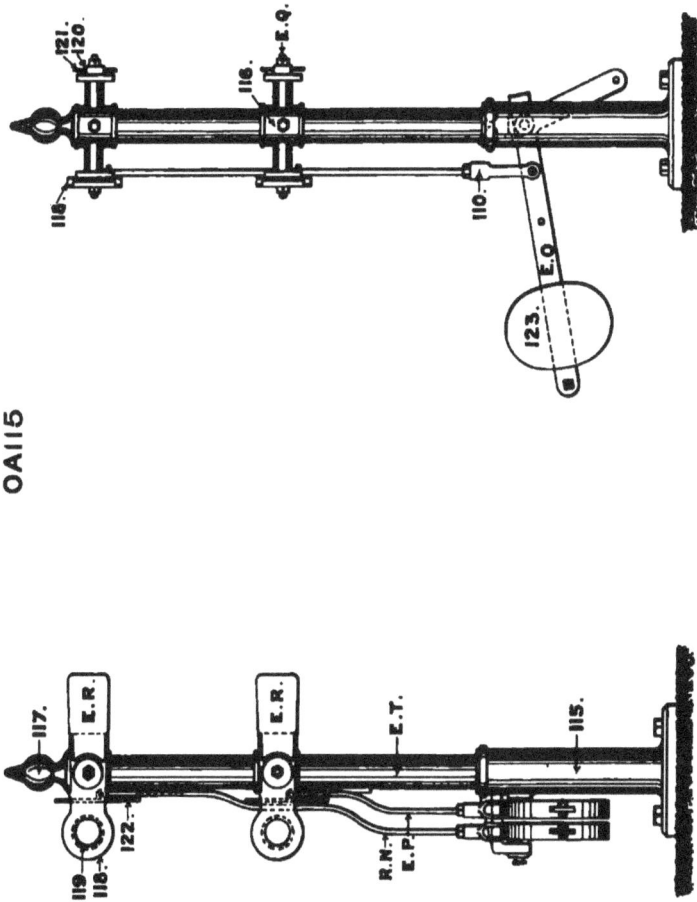

DWARF SIGNALS—Continued

Order No.	
OC115	**Four Arm Dwarf Signal,** complete, with stand and lag screws, **7** feet 9 inches of 2½ inch pipe, top bearing, three centre bearings and set screws, four semaphore spindles with nuts and cotters, four semaphore blade grips with four back light spectacles, eight rings, eight glasses and bolts, four ash blades with bolts, four ⅝ inch solid down rods with eye joints 6 feet 3 inches, 4 feet 9 inches, 3 feet 3 inches, and 1 foot 9 inches in length, four ¾ inch screw jaws with nuts, pins and cotters, four balance levers with four weights, pin and bolts, four Standard dwarf signal lamps with four brackets and set screws.

Note. The illustrations show rod connected signals. When worked by two wires a wheel and chain are supplied in addition to parts named.

FITTINGS

115	**Stand** only (when ordering state if for 1, 2, 3 or 4 arm signal).
116	**Centre Bearing** only.
117	**Top Bearing** only.
ES	2½ inch **Pipe** only, for **One Arm Dwarf Signal.**
ET	2½ inch **Pipe** only, for **Two Arm Dwarf Signal.**
EU	2½ inch **Pipe** only, for **Three Arm Dwarf Signal.**
EV	2½ inch **Pipe** only, for **Four Arm Dwarf Signal.**
118	**Semaphore Blade Grip** only.
119	**Ring** only, for Blade Grips.
TC	3¾ inch **Ruby Glass** only, for Blade Grip.
120	**Back Light Spectacle** only.
121	**Ring** only, for Back Light Spectacle.
TD	3 inch **Blue Glass** only, for Back Light Spectacle.
EQ	**Semaphore Spindle** only.
ER	**Ash Blade** only.
122	**Lamp Bracket** only.
411	**Standard Dwarf Signal Lamp** only.
EP	⅝ inch **Down Rod with Eye Joint** only, for **One Arm Dwarf Signal.**

FITTINGS FOR DWARF SIGNALS—Continued

Order No.	
RN	⅜ inch Down Rod with Eye Joint only, for additional arm for **Two** Arm Dwarf Signal.
RO	⅜ inch Down Rod with Eye Joint only, for additional arm for **Three** Arm Dwarf Signal.
RP	⅜ inch Down Rod with Eye Joint only, for additional arm for **Four** Arm Dwarf Signal.
110	¾ inch Screw for Jaw only.
EO	Balance Lever only.
123	Balance Weight only.
0218	4 inch Wheel for One Arm Dwarf Signal, complete, with stand, wheel, pin, cotters and lag screws.
0138	4 inch Wheel for Two Arm Dwarf Signal, complete, with stand, two wheels, washer, pin, cotters and lag screws.
0139	4 inch Wheel for Three Arm Dwarf Signal, complete, with stand, three wheels, two washers, pin, cotters and lag screws.
0140	4 inch Wheel for Four Arm Dwarf Signal, complete, with stand, four wheels, three washers, pin, cotters and lag screws. Illustrated on page 76.
218	1-way 4 inch Wheel Stand only.
138	2-way 4 inch Wheel Stand only.
139	3-way 4 inch Wheel Stand only.
140	4-way 4 inch Wheel Stand only.
124	4-inch Wheel only.

NOTE.—We now make our Dwarf Signal Blade Grip not as shown in our illustrations but the same form as our Standard C Pattern Blade Grip, because by setting the weight out of the line of the blade, said weight acts more certainly to carry the blade from its safety position to that of danger, than when said weight is in the line of the blade. This is of most importance when the blade accidentally becomes detached from the connections.

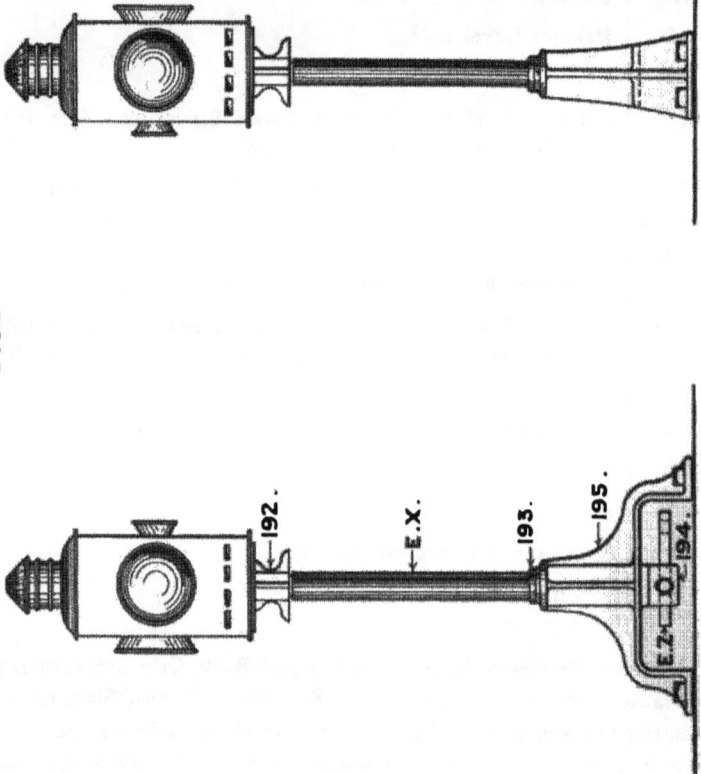

PIPE POT SIGNAL
0195

192.

E.X.

193.

195.

194.

E.X.

COMPENSATING POT SIGNAL
O189

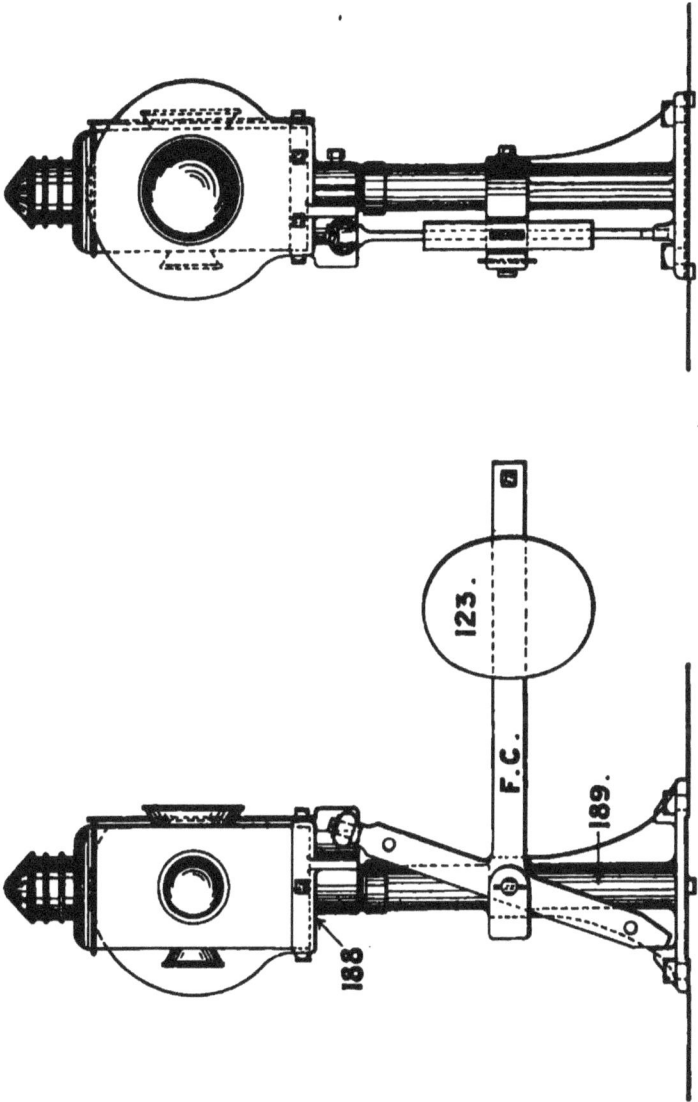

123.

F.C.

189.

188

PIPE POT SIGNAL

Illustrated on Page 108

Order No.	
0195	1¼ inch Pipe Pot Signal, complete, with stand and lag screws, top bearing bush, 2 feet 9 inches of pipe, lamp bracket with standard Pot Signal lamp, crank bush with set screw and adjustable crank.

FITTINGS

195	Stand only.
193	Top Bearing Bush only.
EX	1½ inch Pipe only, 2 feet 9 inches high.
192	Lamp Bracket only.
412	Standard Pot Signal Lamp only.
194	Crank Bush with set screw only.
EZ	Adjustable Crank only.

COMPENSATING POT SIGNAL

Illustrated on Page 109

0189	Compensating Pot Signal, complete, with stand and lag screws, lamp bracket with bolts, standard Pot Signal lamp with two signal discs, balance lever with weight, pin, cotter and bolt, revolving shaft with set screws.

FITTINGS

189	Stand only.
188	Lamp Bracket only.
FB	Revolving Shaft only.
FC	Balance Lever only.
123	Balance Weight only.
412	Standard Pot Signal Lamp only.
TE	Danger Signal Disc only.
TF	All-right Signal Disc only.

LAMPS AND LENSES

Illustrated on Pages 112 and 113

Our Lamps have all the latest improvements for ventilation, removing parts for cleaning, or replacing broken parts, etc. We use a good no-chimney burner, which we can supply singly or by the dozen. Also lenses of the following colors, and all Standard diameters, viz.: Red, White, Green, Purple and Blue. Our Combination Lense or Glass gives the best and only distinct blue light extant.

Order No.

410   **Standard C Pattern Signal Lamp.** This is the Standard Main Line Semaphore Lamp. See pages 78, 81, 83, 84, 85, 87, 89, 90, 92, 94 and 96.

422   **Two Arm Station Signal Lamp.** This Lamp answers for both blades on the Signal. See pages 79 and 80.

411   **Dwarf Signal Lamp,** with small lense to suit Dwarf Signals. See pages 103, 104, 105 and 106.

412   **Pot Signal Lamp.** To fit the Pot Signal stands. See pages 108 109 and 110.

413   **Indicator Lamp.** To fit the Switch Indicator.

414   **Reflecting Lamp.** This Lamp is used for throwing light on to the semaphore blade.

415   **Changeable Semaphore Lamp,** with adjustable Back Spectacle.

423   **Tower Lamp.** This Lamp is calculated to shed most of its light upon the Machine and Operator's table and does not interfere with sight from Tower, nor is it seen by Engine men and so can't be mistaken for a Signal.

LAMPS

410

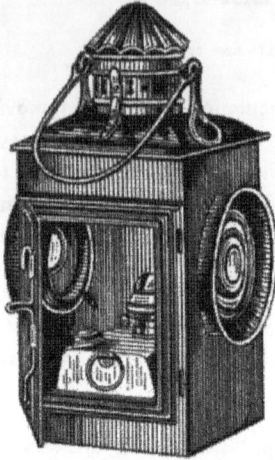

411

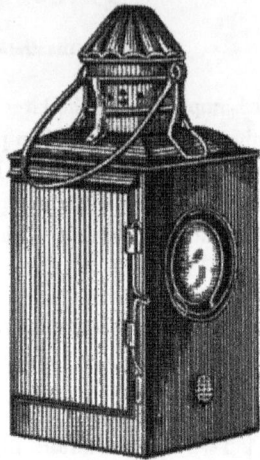

412

413

LAMPS

414

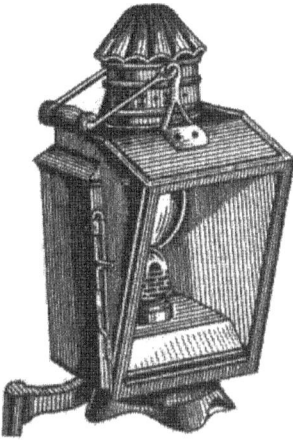

415

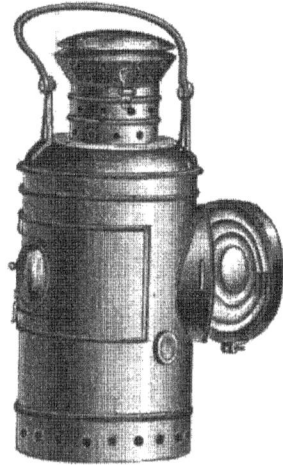

423

SIGNAL TOWER
STYLE A

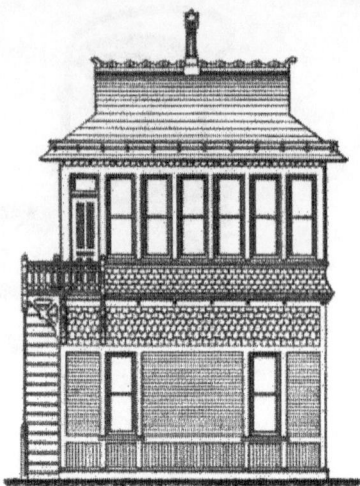

REAR ELEVATION.

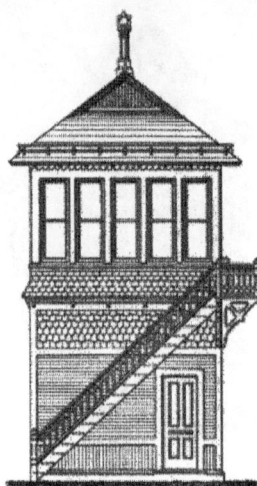

END ELEVATION.

STYLE B

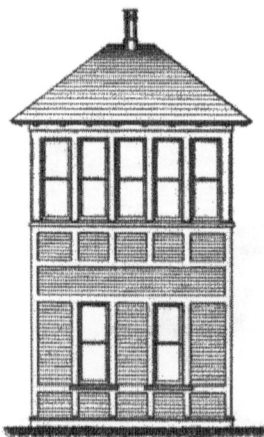

FRONT ELEVATION.

END ELEVATION.

STANDARD SIGNAL TOWERS

SIGNAL TOWER, STYLE A

Our Illustration shows a Tower designed by the Architect of the N. Y. L. E. & W. R. R., 20 feet 1½ inches long, 12 feet wide, 13 feet high from rail to floor and 10 feet high from floor to ceiling and capable of accommodating a machine of 28 levers.

The width of Towers given here is for the Johnson Patent Machines and never varies with the increase of the number of levers, the length only varies, and these variations are shown in a table appended. Should the use of a Saxby and Farmer or any other class of Machine be intended, it may be necessary to increase the width as well as the length. The proportions, projections and ornamentations of this Tower with its handsome outside stairs and landing stage, give it a very effective appearance and make it a desirable tower for positions where the surroundings call for something artistic.

SIGNAL TOWER, STYLE B

Is a modification of Style A, sizes being the same in both, in which the ornamentation has been reduced and the stairs placed inside to save expense. We make also a Tower, Style C (not illustrated), with flat roof and still further modifications reducing expense.

We do not supply Stove, Locker and Cupboard for Towers unless specially mentioned, but trim the roof and ceiling and furnish terra-cotta pipe for smoke pipe to pass through and make good to same.

TABLE OF SIZES

LEVERS	LENGTH OF TOWER		LEVERS	LENGTH OF TOWER	
	FEET.	INCHES.		FEET.	INCHES.
8 and under	12	0	60	33	10
12 to 16	14	8½	64 to 68	36	6½
20 to 24	17	5	72 to 76	39	3
28	20	1½	80	41	11½
32 to 36	22	10	84 to 88	44	8
40	25	6½	92	47	4½
44 to 48	28	5	96 to 100	50	1
52 to 56	31	1½			

TIMBER FOUNDATIONS

Illustrated on Pages 118 and 119

We make all our foundations of white oak or yellow pine, accurately tenoned and morticed by special machinery.

Pipe Carrier Foundations are made for any number of ways, of 2¼ inches by 8 inches white oak or yellow pine, complete, with bolts and washers.

Order No.	
RS	1-way Pipe Carrier Foundations, Length of top, 11¼ inches. See Illustration, page 118.
RT	2-way Pipe Carrier Foundations, Length of top, 1 foot, 2 inches. See Illustration, page 118.
RU	3-way Pipe Carrier Foundations, Length of top, 1 foot, 5 inches. See Illustration, page 118.
RV	4-way Pipe Carrier Foundations, Length of top, 1 foot 7¼ inches.
RW	5-way Pipe Carrier Foundations, Length of top, 1 foot 10 inches.
RX	6-way Pipe Carrier Foundations, Length of top, 2 feet 1 inch.
RY	7-way Pipe Carrier Foundations, Length of top, 2 feet 4 inches.
RZ	8-way Pipe Carrier Foundations, Length of top, 2 feet 6¼ inches.
SA	9-way Pipe Carrier Foundations, Length of top, 2 feet 9 inches.
SB	10-way Pipe Carrier Foundations, Length of top, 3 feet.

(Larger Sizes to Order.)

SC	Style A, Small Crank Foundation, of 5 inches x 12 inches, white oak, complete, with bottom braces, bolts and washers. See illustration, page 118.
SD	Style A, Large Crank Foundation, same as above with 5 feet top.
SE	Style B, Small Crank Foundations, of 5 inches x 12 inches, white oak, complete, with side braces, bolts and washers. See illustration, page 118.

TIMBER FOUNDATIONS—Continued

Order No.	
SF	**Style B, Large Crank Foundations,** same as above with 5 feet top.
SG	**Pipe Compensator Foundations,** of 5 inches x 12 inches, white oak, complete, with side braces, bolts and washers. See illustration, page 118.
SH	**Wire Compensator Foundations,** with top (length to order) of 3 inches x 10 inches, white oak, complete, with two No. 1 Wheel foundations, bolts and washers. See illustration, page 119.
SJ	**No. 1 Wheel Foundations,** of 2½ inches x 10 inches, white oak, complete, with side braces, bolts and washers. See illustration, page 119.
SK	**No. 2 Wheel Foundations,** of 2½ inches x 10 inches, white oak, complete, with side braces, bolts and washers. See illustration, page 119.
SL	**2-way Selector Foundations,** of 2½ inches x 12 inches, white oak, complete, with two 3-way Pipe Carrier Foundations, bolts and washers. See illustration, page 119.
SM	**3 and 4-way Selector Foundations,** of 2½ inches x 16 inches, white oak, top 6 feet in length, 1 foot 4 inches wide, complete, with two 5-way Pipe Carrier Foundations, bolts and washers.

TIMBER FOUNDATIONS

RS   RT   RU   SE

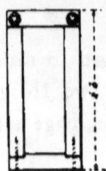 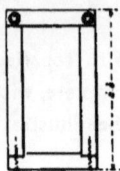 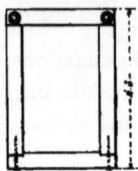 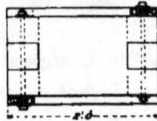

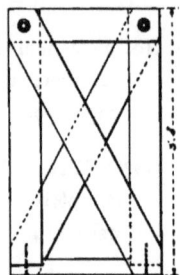

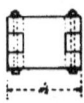 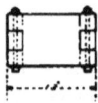 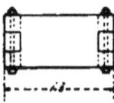

SC

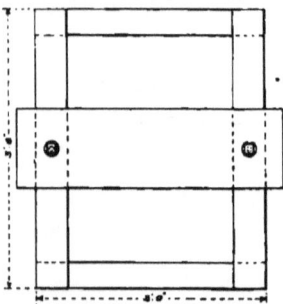

SG

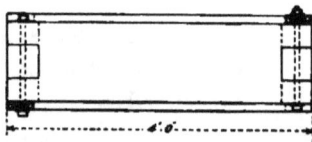

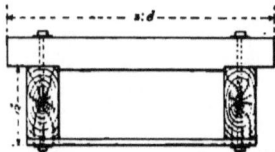

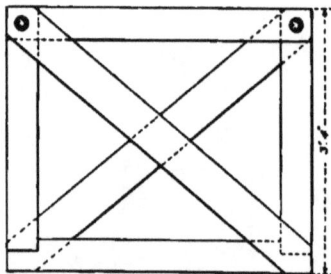

TIMBER FOUNDATIONS
SH

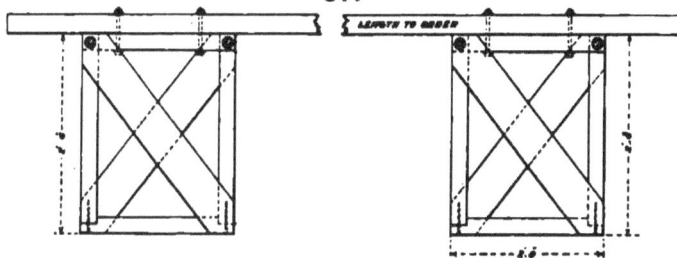

SJ

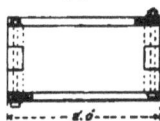

SK

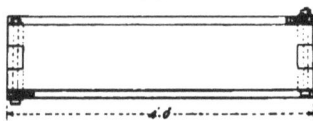

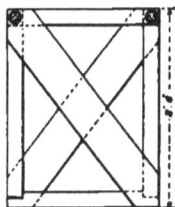

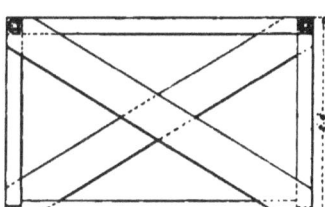

SL

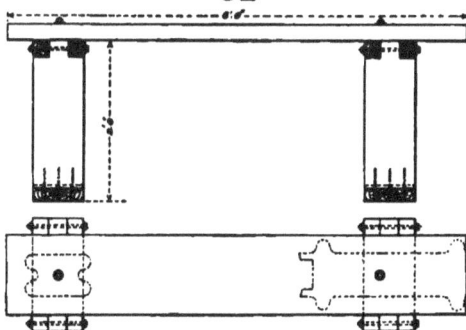

THROW-OVER LEVER SWITCH STAND

PLAN

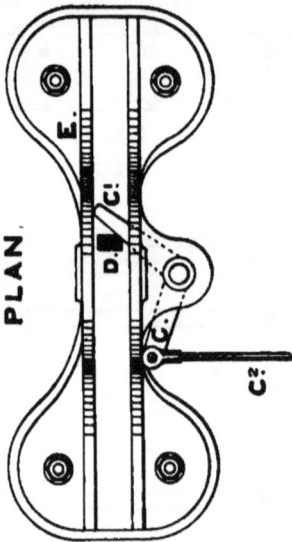

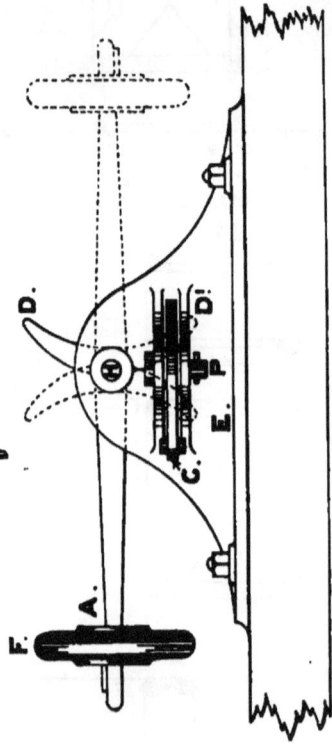

SIDE ELEVATION

THROW-OVER LEVER SWITCH STAND

Our Illustration shows a Throw-over Lever Switch Stand which we recommend to the notice of Railroad Officers, being very simple, durable and cheap.

The lever A throws parallel to the rail presenting the broad sides of the
weight F to the view of the Engine men, one side being painted red and
the other white, indicating whether the switch is set for the Side or the Main
track.   When desired a lamp is mounted on the centre of the Bell Crank C,
giving indications for night.

This Throw-over Lever Switch Stand permits of engines or trains running
through the switches in a trailing direction without injury to the switch or stand.

EXPLANATION OF ILLUSTRATION

E is a stand having lug bearings C^1 which carry the bell crank centre P,
and bell crank C. A is a lever mounted on the centre H and having curved
arms DD^1.   The bell crank C has one arm C^1 which lies across the path of the
curved arms DD^1 and to the other arm C^2, the connecting rod to switch is
jointed.

When the lever A is thrown over from the position indicated by full lines
to that indicated by dotted lines, the arm DD^1 takes up the position indicated
by dotted lines and not lettered, and carries with it the arms C^1 C^2 and throws
the switch.

ROCKING SHAFT BRIDGE COUPLER

Patent applied for

Illustrated on pages 124 and 125

The Illustration shows one of our Bridge Couplers by which signals and switches adjacent to a Draw-bridge are worked from the Draw by the Bridge tender as efficiently as when there are no breaks in the connections.

We have matured an arrangement possessing none of the weaknesses which characterized couplers preceding this and which we believe will give the greatest satisfaction to all using it. In its multiplex form it requires comparatively small lateral space. It is so arranged that before the road can be broken by the swinging of the bridge it is necessary to withdraw that part of the coupler having its bearing on the Draw some inches from contact with that on the abutment, thereby making certain of a perfect clearance of the two ends of the coupler, preparatory to the movement of the Draw, and when the Draw is replaced in its normal position, for Railroad traffic, and not until then, the Draw end of the coupler can be shot forward again into connection with its other part on the abutment, after which, by the movement of a suitable lever or levers, the signals, or switches, or both may be moved according to requirements. For a fuller understanding of the mechanism of the Coupler we refer you to the following description and the illustrations.

Fig. 1 is lay out for one Coupler and its connections.

Fig. 4 is an elevation of the Coupler.

Fig. 5 is a plan of the Coupler.

To operate switches or signals the couplers are made to rotate, thereby communicating motion to ordinary bell cranks and connections. When the two members of the coupler are disconnected they are locked and prevented from rotating as shown in Figs. 6 and 7, and thereby hold the switches and signals which they control in their normal positions. Longitudinal expansion or contraction of the bridge or approach does not impair the efficiency of the coupler, this being compensated for such in the clutch of the coupler, and any deflection or elevation of either end of the coupler is neutralized when they are connected by the spring bearings shown in Fig. 8.

ROCKING SHAFT BRIDGE COUPLER—Continued

Order No.	
0489	**Rocking Shaft Bridge Coupler,** complete, with two stands (for Bridge) and lag screws, two bearings with pins and springs, revolving shaft, swivel, wide jaw, wrought crank, $1\frac{1}{4}$ inch jaw with pins and cotters, two stands (for Abutment) with caps, bolts and lag screws, revolving shaft with key, spring and collars, wrought crank, $1\frac{1}{4}$ inch jaw, pin and cotters.

FITTINGS

485	**Stand** only, for Bridge.
486	**Bearings** only.
487	**Stand** only, for Abutment.
488	**Cap** only.
TJ	**Revolving Shaft** only, with tenon end.
TN	**Revolving Shaft** only, with mortice end.
TH	**Swivel** only.
TR	**Collar** only, for Shaft.
TP	**Key** only, for Shaft.
TQ	**Spring** only, for Shaft.
TL	**Pin** only, for Bearings.
TM	**Spring** only, for Bearings.
TK	**Wrought Cranks** only.
TG	**Wide Jaw** only.
AY	$1\frac{1}{4}$ **inch Jaw** only.

LAY-OUT FOR ROCKING SHAFT BRIDGE COUPLER

Patent Applied For

FIG. 1

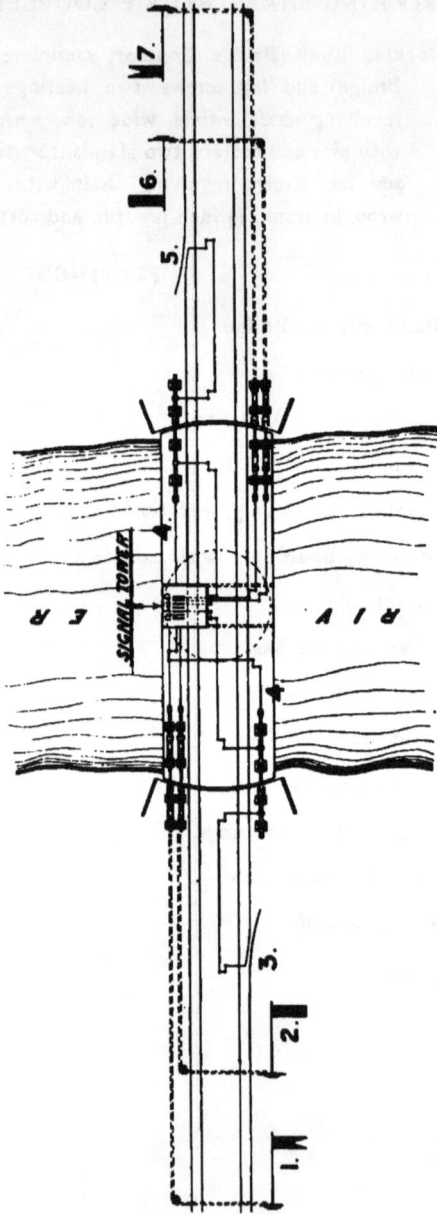

FITTINGS FOR
ROCKING SHAFT BRIDGE COUPLER
Patent Applied For
O489

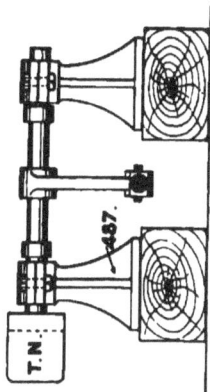

FIG. 4

ABUTMENT

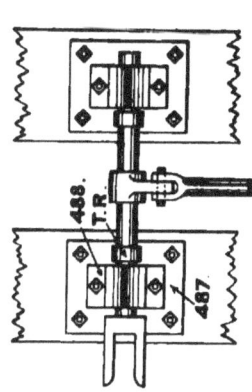

FIG. 5

BRIDGE

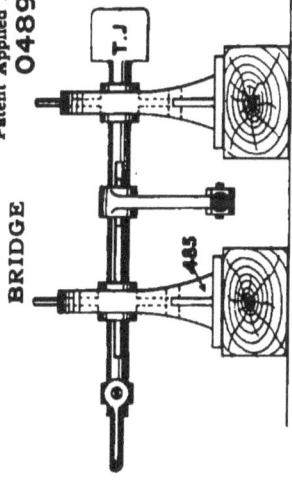

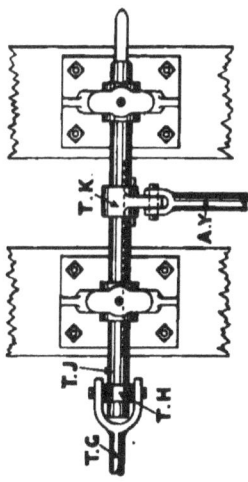

FIG. 7

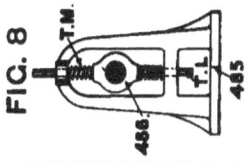

FIG. 8

FIG. 6

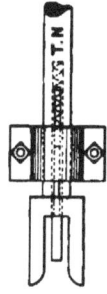

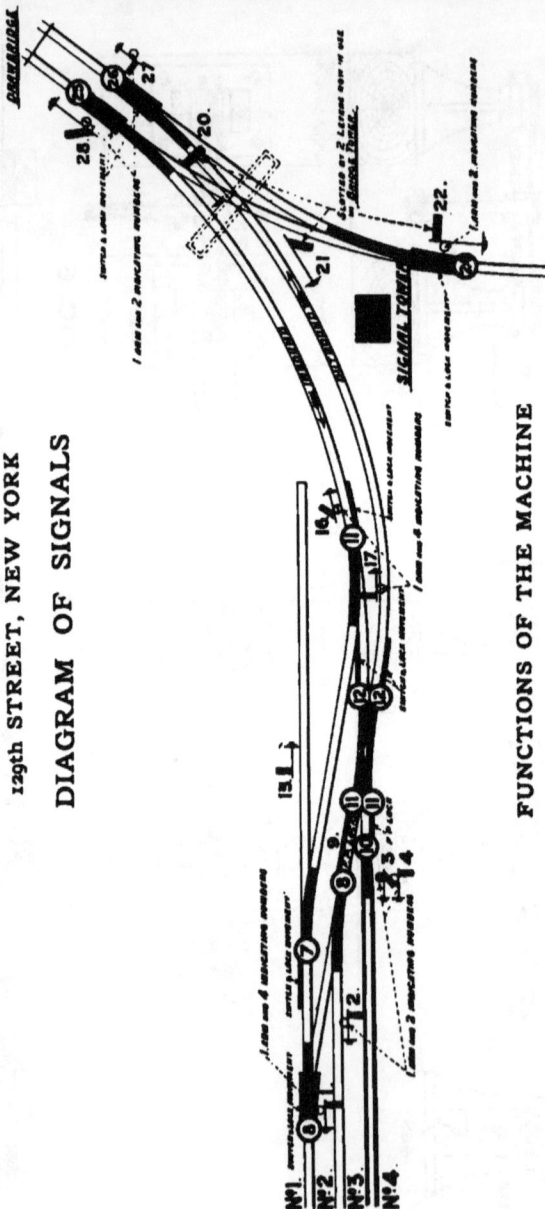

THE SUBURBAN RAPID TRANSIT COMPANY

129th STREET, NEW YORK

DIAGRAM OF SIGNALS

FUNCTIONS OF THE MACHINE

LEVERS

1. Works 1 Signal with 4 Indicators
2. Works 1 Signal with 3 Indicators
3. Works 1 Signal with 3 Indicators
4. Works 1 Signal with 3 Indicators
7. Works 1 Switch, 2 Switch Lock and 1 Detector Bar
8. Works 2 Switches, 1 Switch Lock and 2 Detector Bars
9. Works 4 Switch Locks and 2 Detector Bars
10. Works 1 Switch
11. Works 3 Switches, 1 Lock and 1 Detector Bar
13. Works 3 Switches, 2 Switch Locks and 2 Detector Bars
14. Works 1 Signal
16. Works 1 Signal with 4 Indicators
17. Works 1 Signal with 4 Indicators
20. Works 1 Signal slotted from Bridge Tower
21. Works 1 Signal slotted from Bridge Tower
23. Works 1 Signal with 3 Indicators, slotted from Bridge Tower
24. Works 1 Switch, 1 Switch Lock and 2 Detector Bars
25. Works 1 Switch, 1 Switch Lock and 2 Detector Bars
26. Works 1 Switch, 1 Switch Lock and 2 Detector Bars
27. Works 1 Signal with 2 Indicators
28. Works 1 Signal with 2 Indicators

SIGNALING

AN EXCEPTIONAL SYSTEM OF SWITCHES

At 129th Street, New York

The Illustration shows suitable interlocking at the Junction of two busy lines on the Rapid Transit Railroad at 129th Street, New York. Quick manipulation of switches and signals is of great importance at such a place, and to avoid derailing of trains in their shifting movements, which when they happen are apt to cause obstruction and delay to all or part of main running traffic, the trailing as well as facing switches are provided with detector bars which prevent the reversing of switches when a train is passing over, or standing on them. This would, under the old and yet prevailing system of a separate lever for each detector bar, add considerably to the number of levers and seriously to the number of movements on the part of the switch operator, and consequently retard the work of shifting and passing traffic over the system. But lately we have introduced mechanism by which it is easy and safe to operate the detector bars by the same lever which throws the switches, and although our mechanism is new the system has been in operation on one important railroad many years, but until lately has not appeared to meet with general approval. This probably is due to the fact that until recently it has been open to the objection of not offering the same security as that offered by a separate lever for the detector bar, but this objection has been removed by the introduction of mechanism by which the signals are prevented from being lowered if by any failure in connections the switches do not answer to the movement of their related levers.

This system which we have named the "Switch and Lock Movement" not only reduces levers, but also connections. Our selector, by which several signals are manipulated by one lever, acting through and controlled by the switches, has also the two fold advantage, viz., saving of levers and connections. The

SIGNALING AT 129TH STREET, NEW YORK—Continued

saving of connections is not only an advantage in reducing the cost, but in places such as our Illustration refers to, where space along side the tracks is very limited, a further and important advantage is obtained, viz., suiting the connections to the space available.

A semaphore arm and co-acting numbers indicating each separate route are used where the tracks diverge. Stop Signals cover the foulings, and those giving right of way over the Draw-bridge are controlled or slotted from the Bridge Signal Tower, so that to give a Clear Signal in these cases the joint action of both signal men is required, but a signal may be thrown to danger by the independent action of either man. The locking is nearly all special or conditional owing to the complexity of traffic movements and the nature of the signaling, and could not be satisfactorily performed by any lever machine except the " Johnson."

The Illustration shows the track set and signals at clear from track 3 to Outbound, and from Inbound to track 1.

SIMPLE GRADE CROSSINGS
DIAGRAM OF SIGNALS

PLAN 1

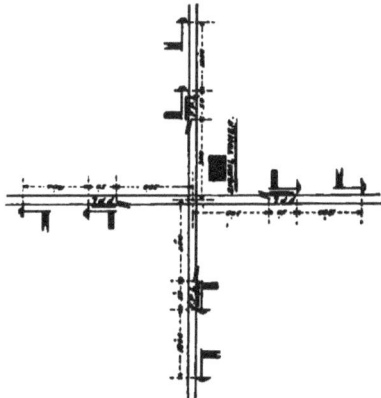

MACHINE

2 Levers for 4 Switches and 4 Locks
8 " " 8 Signals

10 Working Levers

PLAN 2

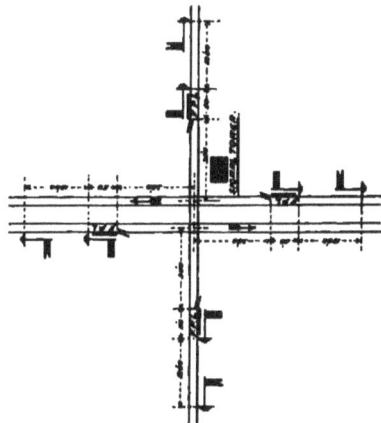

MACHINE

2 Levers for 4 Switches and 4 Locks
8 " " 8 Signals

10 Working Levers

PLAN 3

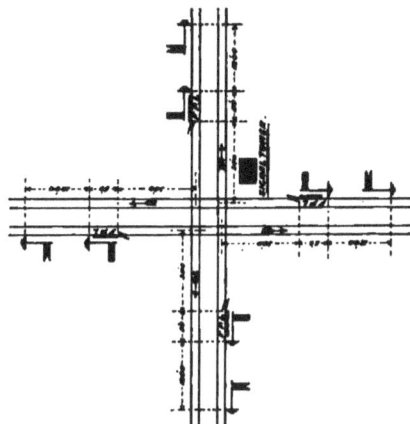

MACHINE

2 Levers for 4 Switches and 4 Locks
8 " " 8 Signals

10 Working Levers

PINS

We manufacture every description of pins with square or turned heads and are all accurately turned to standard gauges of the various sizes, required in connection with signal fittings.

When ordering other than standard sizes, give diameter and distance from underside of head or between cotter holes. We have improved machinery for making every class of special pins.

ORDER NO.	SIZE INCHES		WHERE USED	NUMBER	SIZE INCHES
			PINS	**COTTERS**	
HG	$\frac{3}{8}$ x $1\frac{1}{8}$	CR	Rocker Jaw and Roller .	With two Cotters	$\frac{1}{8}$ x 1
HH	$\frac{3}{8}$ x $1\frac{1}{8}$	CR	Latch Handle . . .	With one Cotter	$\frac{1}{8}$ x 1
HK	$\frac{3}{8}$ x $2\frac{1}{16}$	CR	Tappet Jaw . . .	With two Cotters	$\frac{1}{8}$ x 1
HL	$\frac{3}{8}$ x $1\frac{3}{4}$	CR	Rocker Coupling .	With two Cotters	$\frac{1}{8}$ x 1
HM	1 x $3\frac{1}{4}$	WI	Rocker Centre .	With two Cotters	$\frac{3}{16}$ x $1\frac{1}{2}$
HN	$1\frac{1}{4}$ x 5	WI	Lever Centre .	. . . . .	. . .
HO	$1\frac{1}{4}$ x $6\frac{1}{4}$	WI	End Lever Centre .	. . . . .	. . .
HP	$1\frac{1}{4}$ x $2\frac{7}{8}$	WI	Draught Lever . .	With one set Screw	$\frac{1}{8}$ x 1
HQ	1 x 6	WI	2-way Crank Stand .	With one Cotter	$\frac{1}{4}$ x 3
HR	$1\frac{1}{4}$ x $8\frac{11}{16}$	WI	3-way Crank Stand . . .	With one Cotter	$\frac{1}{4}$ x $1\frac{3}{4}$
HS	1 x $4\frac{1}{4}$	WI	1, 4, 6 and 10-ways Crank Stand .	With one Cotter	$\frac{1}{4}$ x $3\frac{1}{4}$
HT	$\frac{7}{8}$ x $2\frac{1}{4}$	WI	1¼ inch Jaw, Double Jaw and Shackle	With one Cotter	$\frac{3}{16}$ x $1\frac{3}{4}$
HV	$\frac{7}{8}$ x $3\frac{3}{8}$	WI	1¼ Wide Jaw	With one Cotter	$\frac{3}{16}$ x $1\frac{1}{2}$
HU	1 x $4\frac{1}{8}$	WI	1-way Vertical Crank .	With two Cotters	$\frac{3}{16}$ x $1\frac{3}{4}$
HW	1 x $9\frac{1}{8}$	WI	2-way Vertical Crank .	With two Cotters	$\frac{3}{16}$ x $1\frac{3}{4}$
HX	1 x $14\frac{1}{2}$	WI	3-way Vertical Crank .	With two Cotters	$\frac{3}{16}$ x $1\frac{3}{4}$
HZ	1 x $19\frac{1}{2}$	WI	4-way Vertical Crank . .	With two Cotters	$\frac{3}{16}$ x $1\frac{3}{4}$
JA	$1\frac{1}{4}$ x $4\frac{1}{4}$	WI	1-way Lazy Jack Compensator .	With one Cotter	$\frac{1}{4}$ x 2
JB	$1\frac{1}{4}$ x $6\frac{1}{4}$	WI	2-way Lazy Jack Compensator .	With one Cotter	$\frac{1}{4}$ x 2
JC	$1\frac{1}{4}$ x $9\frac{1}{4}$	WI	3-way Lazy Jack Compensator	With one Cotter	$\frac{1}{4}$ x 2
JD	$\frac{3}{4}$ x $1\frac{1}{4}$	WI	Coupling for Lazy Jack . . .	Rivetted	
JE	$\frac{7}{8}$ x $3\frac{1}{4}$	WI	1-way Rack and Pinion Compensator	With two Cotters	$\frac{3}{16}$ x $1\frac{3}{4}$
JF	$\frac{7}{8}$ x $6\frac{1}{4}$	WI	2-way Rack and Pinion Compensator .	With two Cotters	$\frac{3}{16}$ x $1\frac{3}{4}$
JH	$\frac{3}{8}$ x 3	CR	Detector Rail Clip	With one Cotter	$\frac{3}{16}$ x $2\frac{1}{4}$
JK	$\frac{3}{8}$ x $1\frac{1}{4}$	CR	Slide Plate .	Rivetted	
JL	1 x $2\frac{7}{8}$	WI	Driving Plate . .	With one Cotter	$\frac{3}{16}$ x $1\frac{1}{4}$
JN	$\frac{3}{4}$ x $3\frac{3}{8}$	WI	Semaphore Eye	With one Cotter	$\frac{3}{16}$ x $1\frac{1}{4}$
JO	$1\frac{1}{16}$ x $1\frac{3}{4}$	CR	$\frac{3}{4}$ inch and 1 inch Screw Jaw .	With one Cotter	$\frac{1}{8}$ x 1

PINS AND COTTERS—Continued

	PINS			COTTERS	
ORDER NO.	SIZE INCHES		WHERE USED	NUMBER	SIZE INCHES
JP	1¼ x 4⅝	WI	1-way Balance Lever Stand and stand for 1-way Signal movement Style B	With one Cotter	¼ x 3
JQ	1¼ x 6⅞	WI	2-way Balance Lever Stand and stand for 2-way Signal movement Style B .	With one Cotter	¼ x 3
JR	1¼ x 9¼	WI	3-way Balance Lever Stand and Stand for 3-way Signal Movement, Style B	With one Cotter	¼ x 3
JS	1 x 3⅞	WI	No. 1 Pin Plate	With one Cotter	₁⁵₆ x 1½
JT	1 x 6¼	WI	No. 2 Pin Plate	With one Cotter	₁⁵₆ x 1½
JW	¾ x 3½	WI	Plunger Casting, F. P. L. . .	With two Cotters	¼ x 1
GO	¾ x 4⅝	WI	1-way Pipe and 1-way Anti-Friction Carriers	With one Cotter	¼ x 1
GP	¾ x 7¼	WI	2-way Pipe and 2-way Anti-Friction Carriers	With one Cotter	¼ x 1
GQ	¾ x 9⅞	WI	3-way Pipe and 3-way Anti-Friction Carriers	With one Cotter	¼ x 1
GR	¾ x 12⅝	WI	4-way Pipe and 4-way Anti-Friction Carriers	With one Cotter	¼ x 1
GS	¾ x 15⅞	WI	5-way Pipe and 5-way Anti-Friction Carriers	With one Cotter	¼ x 1
GT	¾ x 18½	WI	6-way Pipe and 6-way Anti-Friction Carriers	With one Cotter	¼ x 1
GU	¾ x 20⅞	WI	7-way Pipe and 7-way Anti-Friction Carriers	With one Cotter	¼ x 1
GV	¾ x 23⅞	WI	8-way Pipe and 8-way Anti-Friction Carriers	With one Cotter	¼ x 1
GW	¾ x 26⅞	WI	9-way Pipe and 9-way Anti-Friction Carriers	With one Cotter	¼ x 1
GX	¾ x 29¼	WI	10-way Pipe and 10-way Anti-Friction Carriers	With one Cotter	¼ x 1
GY	⅞ x 3⅞	WI	1-way Vertical Wheel Stand .	With two Cotters	¼ x 1
GZ	⅞ x 6¼	WI	2-way Vertical Wheel Stand .	With two Cotters	¼ x 1
HA	⅞ x 8¼	WI	3-way Vertical Wheel Stand .	With two Cotters	¼ x 1
HB	⅞ x 10⅞	WI	4-way Vertical Wheel . . .	With two Cotters	¼ x 1
HC	⅞ x 3½	WI	1-way Horizontal Wheel Stand and Draught Wheel	With two Cotters	¼ x 1
HD	⅞ x 5½	WI	2-way Horizontal Wheel Stand .	With two Cotters	¼ x 1
HE	⅞ x 7⅞	WI	3-way Horizontal Wheel Stand .	With two Cotters	¼ x 1
HF	¾ x 10	WI	4-way Horizontal Wheel Stand	With two Cotters	¼ x 1
KA	⅝ x 4	WI	Wire Compensator . . .	With two Cotters	¼ x 1
KB	¾ x 5½	WI	1-way Selector Roller Carrier .	With two Cotters	¼ x 1
KC	¾ x 7¾	WI	2-way Selector Roller Carrier .	With two Cotters	¼ x 1
KD	¾ x 10	WI	3-way Selector Roller Carrier .	With two Cotters	¼ x 1
JK	¾ x 12¼	WI	4-way Selector Roller Carrier .	With two Cotters	¼ x 1

PINS AND COTTERS—Continued

PINS				COTTERS	
ORDER NO.	SIZE INCHES		WHERE USED	NUMBER	SIZE INCHES
KL	1 x 3 5/8	WI	Wheel, Signal Movement, Style A .	With one Cotter	3/16 x 1 1/2
KM	1 x 5 1/8	WI	Crank Stand, Signal Movement, Style A	With one Cotter	3/16 x 1 1/2
KO	7/8 x 4	WI	Roller Pin, Signal Movement, Style A	With one Cotter	3/16 x 1 1/2
KS	1 x 4 5/8	WI	1-way Crank Stand, Signal Movement, Style B	With one Cotter	1/4 x 3
KT	1 x 6 7/8	WI	2-way Crank Stand, Signal Movement, Style B	With one Cotter	1/4 x 3
KU	1 x 9 1/8	WI	3-way Crank Stand, Signal Movement, Style B	With one Cotter	1/4 x 3
KW	5/8 x 2 1/4	CR	Roller Pin, Signal Movement, Style B	With one Cotter	1/8 x 1
KX	1 x 4 1/4	WI	1-way Balance Lever, Dwarf Signal .	With one Cotter	1/4 x 3
KZ	1 x 6 1/4	WI	2-way Balance Lever, Dwarf Signal .	With one Cotter	1/4 x 3
LA	1 x 8 1/4	WI	3-way Balance Lever. Dwarf Signal .	With one Cotter	1/4 x 3
LB	1 x 10 1/4	WI	4-way Balance Lever, Dwarf Signal .	With one Cotter	1/4 x 3
LC	1 x 12 1/4	WI	5-way Balance Lever, Dwarf Signal .	With one Cotter	1/4 x 3
LD	1 x 14 1/4	WI	6-way Balance Lever, Dwarf Signal .	With one Cotter	1/4 x 3
LE	5/8 x 2	CR	Eye for Dwarf Signal	With one Cotter	1/8 x 1
LF	3/4 x 3 3/4	WI	1-way 4 inch Vertical Wheel Stand .	With two Cotters	3/16 x 1 1/2
LG	3/4 x 5 3/4	WI	2-way 4 inch Vertical Wheel Stand .	With two Cotters	3/16 x 1 1/2
LH	3/4 x 7 3/4	WI	3-way 4 inch Vertical Wheel Stand .	With two Cotters	3/16 x 1 1/2
LJ	3/4 x 9 3/4	WI	4-way 4 inch Vertical Wheel Stand .	With two Cotters	3/16 x 1 1/2
LK	1 1/4 x 4 1/8	WI	P. R. R. 1-way Crank Stand and 18 inch Hor. Compensators . . .	With one Cotter	1/4 x 1 3/4
LL	1 1/4 x 6 11/16	WI	P. R. R. 2-way Crank Stand	With one Cotter	1/4 x 1 3/4
LM	1 1/4 x 8 11/16	WI	P. R. R. 3-way Crank Stand .	With one Cotter	1/4 x 1 3/4
LN	5/8 x 3 3/4	WI	2, 3 and 4-way Signal Wheel .	With two Cotters	1/8 x 1
LO	5/8 x 6	WI	3-way Signal Wheel	With two Cotters	1/8 x 1
LP	5/8 x 8 1/4	WI	4-way Signal Wheel . .	With two Cotters	1/8 x 1
			ROLLERS		
HJ	1 x 1	WI	Rocker.		
JJ	1 x 1	WI	Rail Clip.		
FA	1 5/8 x 1 1/2	WI	Crank, Signal Movements, Style A and B.		
66A	1 1/4 x 1 3/4	CI	Pipe, Anti-Friction and Selector.		
73	1 3/16 x 1 1/2	CI	Plunger Stand.		
SP	5/8 x 1 1/8	WI	Slot.		

NOTE.—C R signifies Cold-rolled Iron, W I signifies Wrought-iron, C I signifies Cast-iron.

SPRING COTTERS

When ordering give diameter and length of cotter from end of loop.

BOLTS

We keep in Stock a large variety of Bolts, with square or hexagon heads.

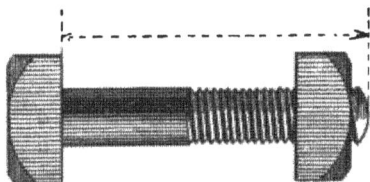

When ordering give diameter, and length from underside of head to end.

LAG SCREWS·

We also keep all sizes and kinds of Lag Screws.

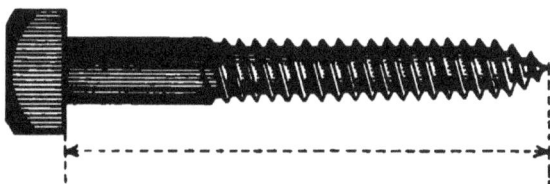

When ordering give length from underside of head to end.

INDEX

INDEX—Continued

INDEX—Continued

INDEX—Continued

INDEX—Continued

INDEX—Continued

www.ingramcontent.com/pod-product-compliance
Lightning Source LLC
Chambersburg PA
CBHW021817190326
41518CB00007B/629